T0401930

Cognitive Systems Monographs

Volume 20

R. Dillmann, Karlsruhe, Germany
Y. Nakamura, Tokyo, Japan
S. Schaal, Los Angeles, USA
D. Vernon, Genoa, Italy

Advisory Board

H. H. Bülthoff, Tübingen, Germany
M. Inaba, Tokyo, Japan
J. A. Scott Kelso, Boca Raton, USA
O. Khatib, CA, USA
Y. Kuniyoshi, Tokyo, Japan
H. G. Okuno, Kyoto, Japan
H. Ritter, Bielefeld, Germany
G. Sandini, Genoa, Italy
B. Siciliano, Naples, Italy
M. Steedman, Edinburgh, Scotland
A. Takanishi, Tokyo, Japan

For further volumes:
http://www.springer.com/series/8354

Davide Rivolta

Prosopagnosia

When All Faces Look the Same

 Springer

Davide Rivolta
Department of Neurophysiology
Max-Planck Institute for Brain Research
Frankfurt am Main
Germany

ISSN 1867-4925   ISSN 1867-4933   (electronic)
ISBN 978-3-642-40783-3   ISBN 978-3-642-40784-0   (eBook)
DOI 10.1007/978-3-642-40784-0
Springer Heidelberg New York Dordrecht London

Library of Congress Control Number: 2013947778

Translation from the Italian language edition: *Prosopagnosia: Un mondo di facce uguali* by Davide
Rivolta, © Edizioni FerrariSinibaldi, an imprint of Sipiss di Ferrari Giuseppe & C. s.n.c Milano (Italy)
2012. All rights reserved
© Springer-Verlag Berlin Heidelberg 2014

Printed on acid-free paper

Springer is part of Springer Science+Business Media (www.springer.com)

Preface

In 2006, after completing my degree in Psychology in Pavia (Italy) I decided to fly (literally) to the other side of the world, to Australia, and join some leading scientists in the very hard task of understanding more about the way our mind works. In April 2007, I started my Ph.D. at Macquarie University, in Sydney. Here I decided to focus my attention toward a specific field, which focuses on understanding the correlates of face recognition both at the behavioral and at the neural level. Prof. Max Coltheart, Associate Professor Mark Williams, Dr. Romina Palermo, and Dr. Laura Schmalzl leaded me toward this challenge that lasted until December 2010, when I completed and submitted my Ph.D. thesis.

After this time spent on face recognition research I was so excited that I wished everyone knew more about this interesting topic. This represents the reason why I decided to write a book that aims to provide a simplified although comprehensive glimpse into the intriguing world of the mechanisms of face recognition. So, I ended up publishing a book in Italian, entitled "Prosopagnosia: Un mondo di facce uguali". This English version does not only represent a translation from the Italian, but, since research on the topic never stops, it also includes very recent results (especially in Chaps. 2 and 3).

What can we read in a face? I bet you have never asked yourself this question. The answer is: "A lot!" In fact, from the face we can retrieve information such as identity, gender, age, attractiveness, race, mood, and approachability of a person. The impressive part is that we can do all of this in a fraction of a second, without even thinking about it (this is probably why you have never asked yourself this question in the first place). Although research has put a lot of effort into trying to understand all those aspects, in this book I will mainly focus my attention on face identification and I try to answer questions like: "Why are humans so fast at recognizing faces?", "Why are we so efficient at recognizing faces?", "Do faces represent a particular category for our visual system?", "Can face recognition fail?".

Of course I am not going to give you the answers now. However, what I can tell you is that in this book I try to summarize and stimulate your curiosity on the line of research that focuses on explaining why humans are generally so good at face processing and why sometimes they are not. I will do this by first providing an introduction to the history, methodology, and techniques commonly adopted for these challenges (Chap. 1). Here, I will not only describe the techniques (there are, in fact, very well done manuals that do it already), but I wish to engage the reader in a wider trip that gives a general idea of what cognitive science is and what it does. After describing a technique, I will provide the reader with some brief descriptions of research that has successfully adopted those techniques. On purpose, I will not provide face-related research since I believe that the reader in this first chapter should have a more general idea of cognitive science and the very different research questions that can be formulated. The expert reader is invited to skip this chapter and directly proceed to Chap. 2; there is no information in Chap. 1 that precludes the understanding of the rest of the book.

In Chap. 2, I will report what I believe is the most relevant research on the cognitive and neural aspects of face processing. In Chap. 3, I will introduce prosopagnosia and in Chap. 4, I describe the intriguing finding of face recognition without awareness. In the last chapter (Chap. 5), I will describe some real cases of people with face recognition difficulties.

Who should read this book? I believe that everyone can read the book. The psychology student can learn something about face processing, a topic that is (to the best of my knowledge) not addressed in great detail in many undergraduate courses. The person interested in science can understand what researchers in the field of cognitive science do, which techniques they use, and what they found. The person who believes to have face recognition problems can use the book to learn something new about their difficulties. Some people say that research often lives in a world parallel to the real one; that is, they never meet. This book represents a modest attempt to make a field of research—the one that focuses on face recognition—available to the general public.

Acknowledgments

Let me use a few words here to thank some people without whom, not only this book, but even all the research I have done so far, would have been not possible. I owe my deepest gratitude to: Prof. Max Coltheart, for the opportunity he gave me to initiate a Ph.D. project very far from my home country, for the support he gave me over my candidature, and for providing insightful feedback on earlier versions of the book; Dr. Romina Palermo, for the constant academic and psychological support, patience (being almost an "Italian mum"), and constructive criticism; A/Prof. Mark Williams for introducing me to the intriguing world of human neuroimaging; Dr. Laura Schmalzl for the precious and broad support provided during the 4 years. Thanks also to Dr. Charlie Stone, Dr. Ellie Wilson, and Isabella Premoli who provided a crucial contribution in the translation of the book and provided insightful feedback on earlier versions of it. Ellie also prepared great part of Chap. 5.

Frankfurt, May 2013

Contents

Chapter 1
Cognitive Science: History, Techniques and Methodology

1.1 The Shaping of Cognitive Science

Among the living creatures on earth, humans represent the only one that can study themselves. Since ancient Greece, philosophers have reflected on many issues including earth's place within the Universe, whether there's an afterlife, the human anatomy, how we remember and perceive the world, the role of science in society, and many others.

Today, most scientists agree that one of the biggest challenges of all times is understanding how the human mind works. It has, step by step, become clear that this challenge cannot be faced without better understanding key structure: the brain. It is now commonly accepted that the mind and brain display a close relationship and, by using an analogy taken from the informatics, the brain essentially acts as the hardware and the mind as the software. Given the importance of the brain-mind relation, different disciplines have contributed in hopes of solving this astonishingly complex puzzle.

Aristotle, Descartes, Locke, Kant and Husserl were amongst the key philosophers who provided very important insights mind–body issue which today represents one of the hottest topics in psychology and cognitive science. Historians, however, share the common consensus that the "birth" of modern, scientific, psychology goes back in 1879, the year in which Wilhelm Wundt setup the first Laboratory of Experimental Psychology in Leipzig, Germany (Hearnshaw 1987; Legrenzi 1997).

Behaviorism

One popular stream of psychological research, particularly in the US, was known as *behaviorism*. Behaviorists such as John B. Watson, Edward L. Thorndike, Edward C. Tolman and Burrhus F. Skinner focused their research on investigating animal behavior (they mainly worked with animals such as dogs, rats and pigeons), their aim being to detail the general *laws of learning* that could be broad enough to be applied to both animals and humans. For an organism, behaviorists conceptualized learning as

D. Rivolta, *Prosopagnosia*, Cognitive Systems Monographs 20,
DOI: 10.1007/978-3-642-40784-0_1, © Springer-Verlag Berlin Heidelberg 2014

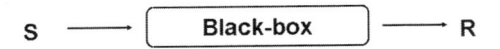

Fig. 1.1 Behaviorists conceptualized the mind as a black-box inaccessible to controlled experiments. Stimuli (*S*) and responses (*R*) represent the only manipulable variables

the ability to apprehend something new, such as an explicit or implicit behavior (e.g., a motor response etc.), that was not known before. They were not interested in the mind or any philosophical speculations about it; they were interested in observable behaviors, which they examined through strict experimental control.

The mind, according to behaviorists, was conceptualized as a *black-box* (see Fig. 1.1), where we can only know what goes in (the stimuli) and what comes out (the response) but we cannot know anything about what happens inside it. According to this view, and by simplifying it a bit, you can only see that there is a relation between looking at a nice plate of Italian pasta and start to salivate, but you cannot understand the processes that mediate it (in other words, you cannot investigate the mind, but only formulate a law that links the stimulus and response).

Cognitivism

It was only in 1967 that Ulric Neisser, an American psychologist, along the same ideas expressed by other scientists including George Miller and Noam Chomsky, coined the term "Cognitivism". Cognitivists believed, in opposition to the behaviorists, that it was indeed possible to *look into the black-box*, and that it was possible examine how information is perceived, transformed, stored and retrieved. In other words, they examined the mind, not just observable behaviors (Ellis and Hunt 1993).

What we know today as *cognitive science* encompasses numerous disciplines including psychology, neuroscience, psychiatry, artificial intelligence, philosophy and linguistics. Each discipline attempts to reveal the secrets of our mind from a different, albeit related, perspectives. Cognitive psychology investigates how the mind works and models its functions. For instance cognitive models of reading describe how it is possible to read aloud and which problems can be related to this ability (Coltheart et al. 2001); cognitive models of calculation describe the steps involved in successfully completing mental calculations and how it can fail (i.e., dyscalculia) (Cipolotti et al. 1991); cognitive models of memory describe how human memory can be distinguished according to different sub-components (e.g., semantic vs. episodic memory), how each sub-component functions and why/when memory does not function properly in some people (i.e., amnesia) (McCarthy et al. 1996). In the same vein, models of face recognition explain the mental steps involved in processing faces and how this process can breakdown (i.e., prosopagnosia) (Bodamer 1947).

Cognitive neuropsychology examines the normal functioning of the mind by investigating (developmental or acquired) cases of brain impairment; *cognitive psychiatry* aims to understand cognitive functioning in individuals with psychiatric disorders such as schizophrenia; and *cognitive neuroscience* tries to understand the neural (physiological) basis of our mind (Caramazza and Coltheart 2006).

In general, the word "cognitive" refers to different human faculties such as attention, perception, memory, calculation and language; the faculties inside the black-box proposed by the behaviorists (Ellis and Hunt 1993; Kandel et al. 2000). In everyday life most of us can focus our attention on what we believe is relevant at any particular moment. For example, we might focus our attention on the traffic light before crossing the road, we can recognize that the object approaching us is a car and that the person driving the car is our uncle Charlie. We can understand language and speak to each other, we can laugh while watching Charlie Chaplin, we can (sometimes) solve very complicated problems and we can play the violin.

In summary, cognitive science aims to understand how all this (and more) works, what the mental steps (stages) involved are, what are their neural correlates, and why, at times, they fail to function properly on a day to day basis as well in those indidivuals who suffer from specific conditions.

1.2 Something About the Brain

Since animals and humans share many neurobiological processes, many researchers attempt to better understand the human mind by conducting research with animals (e.g., mice and monkeys). Despite some issues related to the fact that, of course, a mouse is not a human, many decades of research on animals have provided crucial insights into our understanding the mind-brain relation. In this book I will mainly focus on human research, albeit some reference to animal studies will occasionally be provided (especially in Chap. 2, which reports on the neural aspects of face processing). There have been many techniques successfully implemented in cognitive science over the last 40 years. Here, however, I will focus on describing those techniques most relevant to understanding face recognition research .

Before describing these techniques utilized, the reader needs to have a rudimental knowledge about the hardware: the brain (see Fig. 1.2). The brain is an organ, which weights around 1.6 kg, has a volume of around 1400 cm^3 (you could hold the brain in one hand), and is composed of billions (someone estimates around 30 billion) of cells known as *neurons*. The brain is divided into two asymmetrical hemispheres, left and right and can be broadly divided into four regions named according to the cranial bones on top of them: the occipital lobe, temporal lobe, parietal lobe and frontal lobe (see Fig. 1.2). The surface of the brain is composed of many *circumvolutions* (like wrinkles) allowing for a very large number of neurons in such a restricted space (volume) (Bear et al. 2007; Kandel et al. 2000; Kempler 2005).

While there's much we still do not know about the brain, thanks to the dedication of countless researchers over the last 200 years, we do know some things about it. For example, thanks to the work of Camillo Golgi and Ramòn Cajal, we

Fig. 1.2 This schematic representation depicts the lateral view of the left hemisphere of the human brain. Indicated are the four lobes: frontal (*1*), temporal (*2*), parietal (*3*) and occipital (*4*) (Picture downloaded and modified from http://all-free-download.com/)

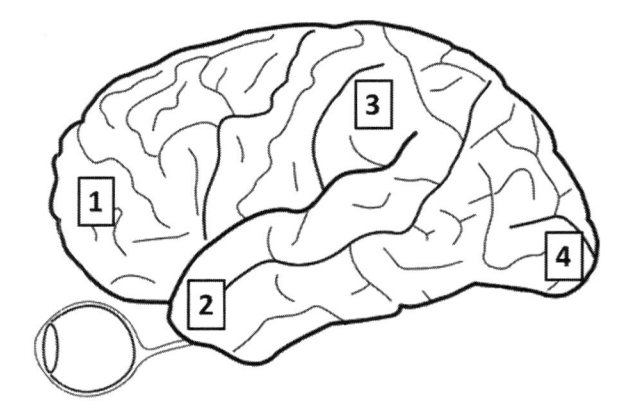

know that neurons constitute the basic components of the brain and that they have physical relations between them called *axons* (you can imagine them as "wires" connecting neurons). For their communication, neurons use electrical impulses called *action potentials* that, by flowing in the axons, mediate the release of *neurotransmitters* contained in specific sacs called *synaptic vesicles* (see Fig. 1.3).

The neurotransmitters are released into the *synaptic cleft* and neurons communicate using *synapses*,[1] whereby the neurotransmitters bind onto the receiving neuron's receptors. The neuron that "sends" the signal (N1 in Fig. 1.3) is called *presynaptic*, the neuron "receiving" the signal (N2 in Fig. 1.3) is called *postsynaptic*.

Today we now know that the idea that different cortical areas of the brain control specific functions, as proposed by the German scientist Joseph F. Gall, the father of *phrenology*, in the nineteenth century, was quite correct. For example, we now know that the frontal lobe enables us to make movements and to control them, to solve problems, to speak and to plan our future; the temporal lobe enables us to hear and to understand language, to recognize faces and objects and to find our way out of the environment; the parietal lobe enables us to perceive hot, cold and pain and mediates the visually-guided physical interaction with the environment; the occipital lobe is essential for us to see our surroundings.

The brain is divided into gray matter, which is the region were neurons are lying, and the white matter, which is made of axons. The cognitive and sensory functions take part in the gray matter (it contains around 10 billion neurons and one million billion synapses!), where action potentials are generated. No action potentials are generated in the white matter. One interesting feature of the brain is that each hemisphere is *contralateral*, that is, sensorial and motor activities come from the opposite side of the body. This means that the very intense sensation of pain you feel in your right hand when inadvertently touching a very hot surface is processed by your left hemisphere in the left *somatosensory* cortex.

---

[1] There are different kinds of synapse, but I am not going into details of synaptic transmission in this book. The interested reader should look at the bibliography provided at the end of the chapter.

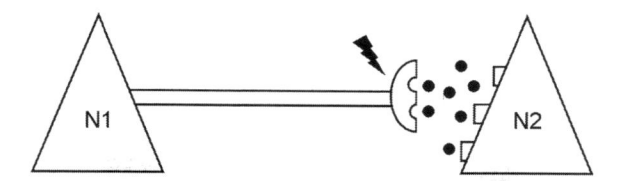

Fig. 1.3 A simplified representation of the basic proprieties involved in the communication between two neurons (here schematized as triangles). *N1* is the neuron that sends the message (the electrical signal called action potential) through the axon, which represents the link between neurons. *N2* is the neuron that receives the message. The space between N1 terminal and *N2* is called the synapse. When the action potential reaches the synaptic button (black lightening in the picture) it signals the release of neurotransmitters (*black dots*). By binding to the *N2* receptors, the neurotransmitter can then excite or inhibit (according to the neurotransmitters involved) *N2*, increasing or decreasing, respectively, the probability to propagate an action potential in *N2*

Up until now, I have focused on providing broad overview of the history of cognitive science and providing basic facts about the brain; however, we still do not know how to "measure" the information processed by the brain. In other words: "How can we investigate the brain and its relation with our mind?" To answer this question, cognitive scientists rely on both behavioural and neuroimaging techniques (Kandel et al. 2000; Kempler 2005; Kutas and Federmeier 1998).

1.3 Behavioural Techniques

In a typical behavioural experiment, the experimenter will manipulate one variable, the *independent variable,* examine how it influences a second variable, the *dependent variable*. The dependent variable reflects some behavioural measure. Let me give you an example. Imagine that as a student, you do not study much, but you are still passing your classes (albeit barely). Your mom, after talking to your teachers, insists that you have great potential, but you are not working hard enough. She insists that you spend more time reading/studying Dante Alighieri's Divina Commedia, your philosophy material and less time playing football with your friends. You are, however, convinced that time spent studying has no relation to the mark you will ultimately receive. To test this, you could do a small experiment: does the number of hours a student spends studying influence the average mark they receive at the end of the year? Here, the number of hours spent studying represents the independent variable and the average mark they receive is the dependent variable.

By interviewing 200 people at your school, and by using some statistical tools you could examine whether there is a statistical significant difference in performance between people that study, say, 3 h a day and people that study 6 h a day. If there is a difference, your mom might be right; however, if there is not any statistically significant difference you can proudly say to your mom that you would rather enjoy time outside than studying in the library.

Let's go back to cognitive science now. Behavioural techniques in cognitive science refer to tasks (typically administered on a computer) that have accuracy and/or reaction times as dependent variables. We will use an example from face recognition research. Imagine that, in a hypothetical experiment, we show a sequence of 50 faces (25 very famous and 25 of completely unknown people) to 100 subjects, each of them individually presented for 1s on a computer screen. We ask participants to press the key "F" when the face is familiar and "U" when the face is not-familiar. We can measure how fast participants are in pressing the button and how accurate they are, thus estimating the average accuracy and the average reaction time for familiar and unfamiliar faces. We may find that participants, on average, are faster in pressing the button when a face is familiar then when it is not, but that accuracy is similar to both categories (please note that these results are imaginary).

Now imagine that you test a new participant which tells you that he cannot follow the plots in movies because he has trouble recognizing the faces of characters in it (for instance he often gets confused with Robert De Niro and Al Pacino's faces). You might find, by using your task, that this subject is much less accurate and much slower than the 100 people you tested. Well, you have just used a behavioural task to tell us something about human behavior (i.e., humans are faster in judging a face as familiar than unfamiliar even though they have the same accuracy in the recognition of both) and you have just "diagnosed" a participant with potential problems in face identification (so you have been a good cognitive neuropsychologist too!).

Interestingly, researchers can now examine behavioral tasks while monitoring their brain activity by using non-invasive neuroimaging techniques. This is useful when the research question focuses on finding the anatomical substrate of a particular cognitive function and/or when the researcher is interested in the timing of cognitive processing (i.e., how long does it take to recognize a familiar face?). I will now discuss the rationale behind some of these neuroimaging techniques and the data we can extract from them.

1.4 Neuroimaging Techniques

Neuroimaging techniques can be divided into structural and functional. On the one hand, structural neuroimaging provides a view, in vivo, of the anatomy of the brain. On the other hand, functional techniques provide a view the brain in action, that is, brain activity. Each technique is used for different purposes. With structural imaging it is possible to see, for instance, whether a clinical population (e.g., people suffering from Alzheimer's disease) have smaller brain volume (gray or white matter) than typical subjects. With functional neuroimaging techniques it is possible to see where and/or when a particular process is taking place in the brain. For example, we can use it to understand which part of the brain we engage to find our way out in unfamiliar environment (i.e., navigation skills).

In this chapter I describe only non-invasive techniques. "Non-invasive" means that no radioactive contrast is intravenously-injected, nor any invasive surgery (i.e., craniotomy) is performed.

1.4.1 Structural Neuroimaging

In the nineteenth century the most common method to examine the neural bases of specific behaviours was by investigating individuals with brain injuries (e.g., strokes) after they passed away. Such studies are known as *post mortem studies*. This means that anatomists had to open the skull, look at the brain, find the location of the brain lesion, and correlate the impaired cognitive function to the site of the brain lesion. It seemed clear that if function A is impaired in a subject and that area A' shows a damage, then A' is likely to represent the neural correlate of A. In this way, Paul Broca (nineteenth century) discovered that problems with speech production (function A) are related to damages to the left hemisphere and, in particular, to the left lateral frontal lobe (lesion A'). This method was the so called *anatomo-clinical correlation*. Fortunately, nowadays we can examine the brain without the need or patience to wait for people to die. Modern science has developed non-invasive techniques to examine both the structure and functions of the brain and its different regions. The first structural technique I will describe is the *Magnetic Resonance Imaging* (MRI).

MRI provides a 3D high-resolution image of the brain. Participants are asked to lie down into a ~60 cm wide bore of a scanner that weighs a few tons (see Fig. 1.4). When safely in the scanner, a magnetic field (commonly around 3.5 Tesla, but it can extend, for human research, until up to 9 Tesla[2]) is applied.

This is the sequence of events that occur during a normal MRI scan:

- The magnetic field generated by the MRI makes the hydrogen (H^+) atoms contained in the water molecule (H_2O) to align to the direction of the magnetic field. Note that H^+ is abundantly present in the brain.
- A radio frequency (the same used for FM radio) is then applied for a few milliseconds.
- This then causes H^+ to be temporary misaligned from the main magnetic field.
- As soon as the radio frequency is stopped, the H^+ atoms go back to their common orientation (the orientation of the applied main magnetic field).
- In doing so, they emit radio frequencies (again, the same FM frequency) such that, after being processed by powerful computers, provides 3D picture of the brain.[3]

---

[2] Tesla (T) is the international unit of measure for the magnetic field. The magnetic field applied inside an MRI scanner is around 100000 times bigger than earth's magnetic field.

[3] In this simplified description of MRI, I voluntary did not mention specific terminology such as Larmor frequency, relaxation time, Fourier transform, gradient field and so on. The interested reader should have a look at the bibliography provided at the end of the chapter.

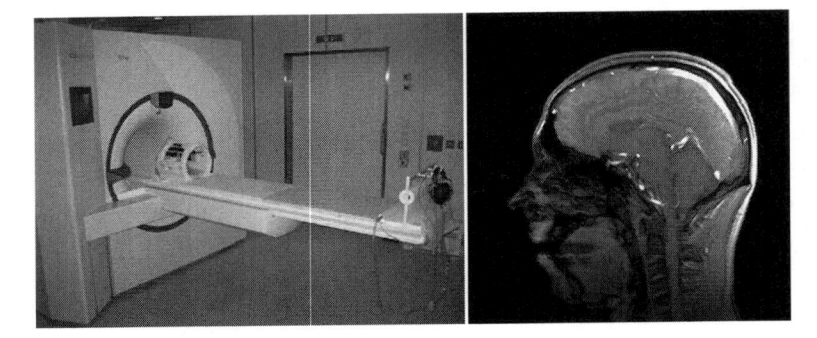

Fig. 1.4 On the *left* is a scanner for MRI/DTI/fMRI (TRIO, Siemens). A typical MRI sagittal slice is on the *right*

The resolution of the obtained image (see Fig. 1.4, right, for an example) is composed of *voxels*, which is a kind of 3D version of the 2D pixels in TVs. The voxel represents the smallest unit of the image. Therefore, the more voxels, the more the image has a better resolution. The brain image obtained from an MRI scan is made of hundreds of thousands of these voxels, each one covering ~55 mm^3 of brain volume. Within this volume there are 5.5 million neurons and 250 km of axons! Thus, when we use MRI, the focus is not on individual neurons, but on thousands of them (Kandel et al. 2000; Logothetis 2008).

Another neuroimaging technique is *Diffusor Tensor Imaging* (DTI) (Hagmann et al. 2006). By adopting the same scanner as the MRI, DTI detects the diffusivity of water molecules inside the white matter to create a 3D image of the fiber-tracts which connect different brain areas. In general, structural imaging is important not only for the understanding of clinical conditions, but even for the investigation of the developing brain (in normal subjects). In fact, researchers can investigate which regions develop faster than others and which regions get pruned (reduce their size) over the course of development.

Recent research has also shown that these structural techniques are becoming popular because of their power in making "prevision". For instance, given the individual brain anatomy, MRI and DTI can predict with high accuracy whether a person is at risk of developing serious pathologies such as Alzheimer or Schizophrenia (Cohen et al. 2011; Koutsouleris et al. 2009). This has very important predictive/diagnostic implications.

1.4.2 Functional Neuroimaging

As discussed above, with structural neuroimaging we can examine the brain anatomy. With functional neuroimaging we can go further and, for instance, examine the neural activity on the brain obtained by an MRI scan. To put it simply: we can see "spots" of activity on the MRI image representing areas of activation for specific tasks/behaviors or conditions.

1.4.2.1 Functional Magnetic Resonance Imaging

One of the functional neuroimaging techniques that has received the most attention over the last 20 years is called functional Magnetic Resonance Imaging (fMRI). With fMRI it is possible to monitor the blood flood within the brain, which, it is believed, reflects neuronal activity. The principle is: the more blood is recruited by a region, the more this region should be active in a particular task or condition. This approach is called BOLD (*blood-oxygenation-level dependent*) and it takes advantage of the magnetic properties of hemoglobin.[4] FMRI provides an indirect measure of brain activity since it focuses on the blood flow and not on the direct neural activity (i.e., action potentials or post-synaptic activity, see Fig. 1.3). By using statistical tools it is possible to ascertain whether "condition A" shows a stronger response than "condition B" in a specific brain region when conducting a particular task (Amaro and Barker 2006; Logothetis 2008; Norman, Polyn, Detre and Haxby 2006). For instance, there is a specific brain region in the medial temporal lobe called the *parahippocampal place area* (or PPA) that responds preferentially to natural and/or artificial pictures of scenes (condition A), such as buildings, lakes and mountains. It does not respond the same to other categories of visual stimuli such as faces (condition B). It is now believed that PPA plays a critical role in the ability of humans to recognize places and in finding our way around known and unknown routes (Epstein and Kanwisher 1998).

Due to its excellent spatial resolution (in the order of millimeters), fMRI is ideal when the research question is mainly focused on knowing "*where*" a certain process of interest is taking place in the human brain.[5] However, there might be situations when it's ideal to know "*when*" something is happening. If a researcher is interested in the *timing of cognition* then fMRI is not ideal as it can only detect slow processes (in the order of seconds). For good time resolution, a different technique is needed such as Electroencephalography (EEG) and Magnetoencephalography (MEG). Both of these techniques have resolutions in order of less than 1 ms! These two techniques have similar, albeit not identical, neurophysiological origin. Contrary to fMRI, EEG and MEG represent direct measures of brain activity since they provide an online index of neural functioning (Kutas and Federmeier 1998; Singh 2006).

1.4.2.2 Electroencephalography

EEG signals are electric signals generated by the brain during behavioural tasks or even at rest, such as during sleep. EEG is recorded from the scalp surface by using electrodes pasted with gel to increase the electric conductance through the skin (see Fig. 1.5). This activity gets amplified and conveyed to a computer that provides elaborate statistics over the electrodes (Bear et al. 2007).

---

[4] The protein in the red blood cells that carries oxygen from the lungs to the rest of the body.

[5] Note that fMRI is even used in animal studies, especially in monkeys.

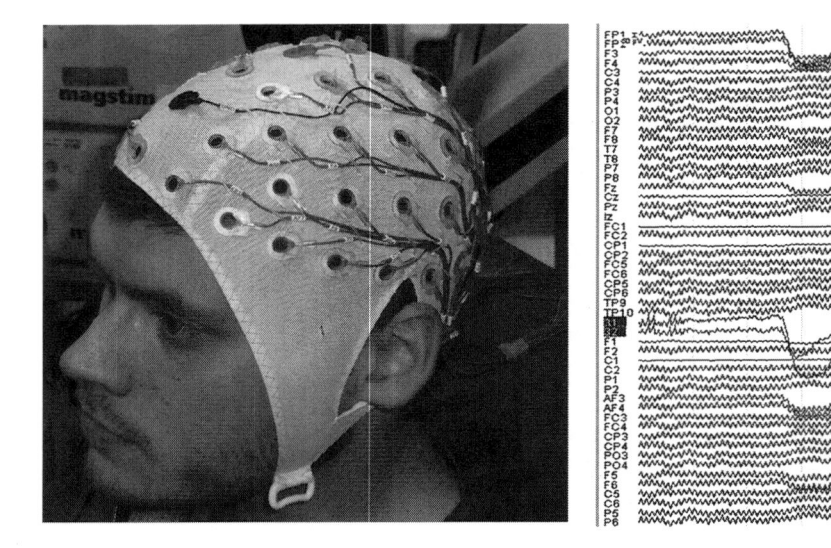

Fig. 1.5 On the *left* is a typical EEG setting, with the cap containing electrodes (in this case 64 electrodes). On the *right* is an example of raw EEG activity as seen at the electrodes level (i.e., before the data analysis)

EEG experiments take place in well isolated laboratories where environmental noise can be excluded, thereby preventing any "contamination" of the data. You might have seen on TV some documentaries on sleep where they discussed the importance of EEG in detecting specific sleep states such as the rapid eyes movement (R.E.M.) state. Today we know that brain functioning is related to rhythmic activity. For example, *alpha* waves (8–12 Hz) are slow waves that characterize the EEG of people when they close their eyes, *delta* waves are very slow waves (<3 Hz) that can be found in deep states of sleep or in disease states such as a coma, and *gamma* waves (30–200 Hz) characterize active cognitive functioning such as memory, perception and attention (Buzsaki and Draguhn 2004; Fries et al. 2007; Tallon-Baudry 2009).

Normally, cognitive scientists are interested in the brain activity associated with completing specific tasks. In cognitive science, EEG is often used to examine brain activity that is *synchronized* with the presentation of a stimulus. The assumption is that, if a specific neural population made of thousands of neurons, responds to a specific stimulus, they will respond in synchrony, that is, they will respond (more or less) at the same time. This synchronized activity gives rise to a strong signal that is detected by EEG by adopting a technique that measures the so called *event related potentials* (ERPs). Let me give you an example. Some researchers were interested in the measurement of some basic, automatic auditory processes using ERPs. They found that, when participants are presented with sequences of auditory stimuli such as "ba, ba, ba, ba, ba, ba", the presence of a deviant stimulus such as "ta" produces a negative ERP component around 150–250 ms post-deviant onset. This is called *mismatch negativity* (as reviewed in Näätänen et al. 2007). An example of a sequence producing the mismatch negativity would be: "ba, ba, ba, ba, ba, ba, ba, ba, <u>ta</u>, ba, ba, ba, ba, ba, ba, ba, ba". Since the mismatch negativity is both present when participants pay attention to the task, and when they don't, for example, while watching a silent movie with subtitles (Peter et al. 2010), it is believed to represent pre-attentive, memory-based auditory discrimination processing or a neural adaptation process. The discovery of this component had great importance as it represents an easy-to-implement paradigm for the investigation of elementary sensory processing in infants, young kids and even clinical populations such as patients with schizophrenia who are unable to complete complicated behavioural tasks. As a note, it has been found that patients with schizophrenia show abnormal mismatch negativity, indicating early sensory processing deficits (see Uhlhaas and Singer 2010 for a review).

1.4.2.3 Magnetoencephalography

MEG, on the other hand, records the tiny magnetic fields generated by the brain at rest or during a task performance. As the reader knows by now, the brain functions using electric signals; those signals are measured by EEG. Flowing current, however, generates a (orthogonal) magnetic field that can be detected by means of special (very sensitive) sensors called SQUIDs (superconducting quantum interference design) contained in a dewar, which is placed in contact to subjects' head (see Fig. 1.6). Since they are superconductors they need to be placed in a very cold environment. This is why SQUIDs are submerged in liquid helium at a temperature close to –270 °C. Since MEG detects magnetic fields that are 1 billion times smaller than the earth magnetic field, the MEG system is built inside a very thick magnetically shielded room that protects it against the external magnetic field.

It is believed that EEG and MEG signals are generated by similar neural generators. However, because of the physics behind the two techniques, they record different aspects of the neural signal (Hamalainen and Ilmoniemi 1994; Singh 2006).

Overall, even to justify its costs (an MEG system can cost ~2 million euros against ~50.000 euros for EEG), MEG is better in localizing neural activity than EEG and it is much better for examining processes that occur very quickly (even above 1000 Hz).[6] Thus, MEG not only has an excellent temporal resolution, but also very good spatial resolution. This makes the MEG the ideal tool for investigating spatio-temporal features of cognitive processes. MEG has also been successfully adopted for the detection of mismatch negativity (Hari et al. 1984). I will provide an example of MEG research in the following chapters as it will be the topic in focus. In summary, with MEG and EEG you can investigate similar problems, but with MEG you can better localize the neural origins of the signal (Gross et al. 2013).

---

[6] Hertz (Hz) is the unit of measure for frequency. By saying that the MEG system can sample at 1000 Hz, it means that in 1s we have 1000 samplings (estimates) of the neuromagnetic activity.

Fig. 1.6 An example of a
MEG scanner. The walls
of the chamber are made
of metals that protect
the system, shown in the
middle of the figure, from
the external magnetic field.
This picture depicts the
275-sensors CTF system used
at the Brain and Imaging
Centre (BIC) in Frankfurt am
Main, Germany

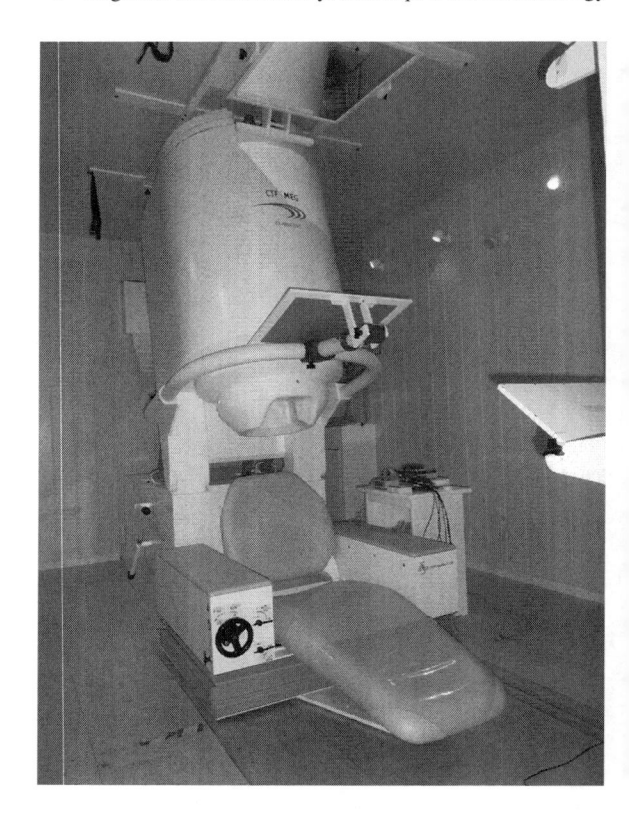

1.4.2.4 Transcranial Magnetic Stimulation

Another technique which uses the magnetic field in order to investigate the
mind-brain relationship is called *Transcranial Magnetic Stimulation* (TMS). It
works thanks to the principle of electromagnetic induction, where the current,
which is flowing in a coil, generates a magnetic field (Fig. 1.7). The coil is placed
above the head, typically on the area that we want to stimulate, like the motor,
sensorial or visual cortex. The main characteristic of TMS is the capacity to detect
a *causal* link between the brain area and its corresponding function. This is not
possible, for instance, with fMRI, where the results have a correlational nature.
In theory, if after inducing a 'virtual lesion' in area A we should notice a deficit in
behaviour A', but not behaviour B'. Therefore, it is possible to establish a causal
relation between A and A'. This relation is even stronger when a virtual lesion at
hypothetical brain area B reduces the performances of B' but not A'. In neuropsychology, the pattern that links A–A' and B–B' is called double dissociation. Here
is an example of a typical application of TMS in cognitive science. FMRI studies demonstrated the existence of an area, located in the occipital lobe, which is
selectively linked to the vision of bodies. It is called the "extrastriate body area"

Fig. 1.7 The figure represents the TMS coil, which is positioned directly on the scalp

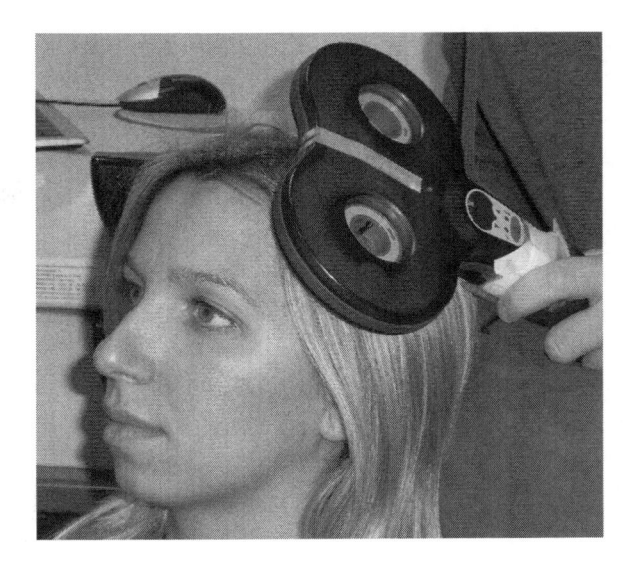

(EBA). EBA is not activated by other categories of visual stimuli, such as objects (Downing, Jiang, Shuman and Kanwisher 2001).

Research with TMS has shown that a virtual lesion at EBA causes a significant reduction in tasks related to body processing, but not of other categories of visual stimuli such as faces and objects (Urgesi et al. 2004). The main advantage of TMS is clear: it creates a specific virtual lesion and allows the researcher to study the subsequent associated (abnormal) behaviour, all this without having to wait for post-mortem analysis (the typical approach of the XIX century).

1.4.2.5 Skin Conductance Response

The last technique I will describe is a measure of the skin conductance response (SCR). Typically, emotional situations involuntarily elicit the release of sweat. This improves the conductance of the skin. In a sense, the skin becomes a better conductor for the current. If a tiny (no-harmful and imperceptible) electric current is applied to the skin surface, the SCR machine detects the resistance of the skin to this current. In other words, when sweat glands release sweat, the resistance of the skin diminishes (that is, the conductance increases) and this can be detected and measured (Dawson et al. 2000).

Interestingly, this physiologic phenomenon has been widely used in the past for "lie detection methods" (Ben-Shakar and Furedy 1990). The idea being, only the culprit of a crime knows details about it (e.g., the crime scene). As such, he/she should (involuntary) increase the SCR when exposed to salient details that only he/she, and not innocent people, should know. The non-controllable (for most people) physiology of a criminal can trick him/her! And though the reliability and efficacy of SCR in police investigations has been called into question numerous times (see Iacono 2011 for a critical investigation on the issue), SCR has been used extensively over the last 40 years by cognitive science researchers. Let me give you a curious example.

Some individuals suffer from a rare psychiatric disease called *Capgras syndrome*. These patients believe that one or more relatives have been substituted by an impostor (or an alien). This impostor, they acknowledge, is physically the same, with the same voice, attitudes and so forth, but is a different person. Of course, this is a delusion and the relatives are actually real, not aliens (so we hope!). An insight into this condition comes from research indicating that normal individuals generate a larger SCR when they view a familiar face compared to an unfamiliar face. Individuals with Capgras delusions do not show the same level of activity when viewing faces of their relatives. This might explain the reason why, even though they believe that the impostors are very similar to the original relatives, *they do not feel it.* It seems that they have no physiological reaction to familiar people, as if they had a silent affective system (Coltheart et al. 1997; Ellis and Lewis 2001).[7]

This chapter does not intend to be a summary of all the techniques available in cognitive science. Rather, merely to provide a general and simplified description of the techniques (behavioural techniques, MRI, DTI, fMRI, EEG, MEG, TMS, SCRs) used in most of the studies on face processing I am going to describe in the next few chapters.

It is worth noting that it has now become more common in the cognitive sciences to use multiple techniques in combination. For instance, it is possible to record EEG while inside the fMRI or MEG scanner, or even record fMRI activity and use it to constrain some spatial properties of MEG data (Singh 2006).

1.5 Single-Case Studies Versus Group Studies

Let me now give you a very quick introduction to an important methodological aspect in cognitive science, that is, the difference between conducting a single case or a group study. Both behavioural and neuroimaging investigations can be done at two different levels. When we have a special, exceptional case it is common practice to conduct a *single-case* study. In single-case studies, a detailed description of the cognitive and neural features of a person are reported and examined. These are then compared to a (usually small) control group, which is made of people with normal intellectual, cognitive and psychological functioning. A good examples are the patients suffering from Capgras syndrome as described above. Since those patients are so rare, the accurate description of their condition tells us a lot about typical cognitive functioning (remember that cognitive

---

[7] Just keep in mind that in Chap. 3 we will discuss again about Capgras Syndrome as possibly representing a complementary condition to prosopagnosia.

neuropsychology aims to study brain disease in order to infer about the normal functioning of the brain).

The main disadvantage of single case studies involves th reliability of the results. Is the behaviour we measured typical of all people with Capgras Syndrome, or it is specific to the patient I am examining? Might other variables I am not taking into account explain the single subject data? Of course, the detailed assessment of further single cases over the years will provide a more solid basis to better understand the common factors involved in the same condition. When researchers examine more common conditions such as Alzheimer's disease, Parkinson, Schizophrenia or people recovering from a stroke, it is common practice to conduct *group studies*. The advantage of group studies is that they provide more information about the general features of a particular condition. However, since small individual variations tend to get cancelled out in group studies, researchers might miss specific details about individual patients which, at the single-case level might have emerged. Group studies can be conducted with 6–15 subjects or even with hundreds/thousands of them. There is no numeric limit and, usually, the more the better since the *power* of an experiment (i.e., the ability to detect an effect if there is one) increases with the number of observations (Shallice 1988). From now on, I will report (and indicate) when the results I describe come from single-case studies. In all other instances, the data are are derived from group-studies

1.6 Conclusions

Hopefully this first chapter has given you a general idea of the brain anatomy and the very basic principles behind its functioning. Again, hopefully you now have a clearer picture of the kind of questions cognitive neuroscientists try to answer every day and the techniques and methodologies at their disposal. You might even have a better idea of the hard work people such as physicists, engineers, psychologists and biologist put into developing and building the techniques described above, in using them, in maintaining them and reporting their results to both the scientific community and the general public. With this information in mind, we can now start to explore research on face recognition.

References

Amaro, E., & Barker, G. J. (2006). Study design in fMRI: Basic principles. *Brain and Cognition, 60,* 220–232.

Bear, M. F., Conners, B. W., & Paradiso, M. A. (2007). *Neuroscience: Exploring the brain* (3rd ed.). In Lippincott Williams & Wilkins (Eds.), Philadelphia, PA, US.

Ben-Shakar, G., & Furedy, J. J. (1990). *Theories and applications in the detection of deception.* New Yourk, US: Springer.

Bodamer, J. (1947). Die Prosop-agnosie. *Archiv fur Psychiatrie und Nervkrankheiten, 179*, 6–53.

Buzsaki, G., & Draguhn, A. (2004). Neuronal oscillations in cortical networks. *Science, 304*, 1926–1929.

Caramazza, A., & Coltheart, M. (2006). Cognitive Neuropsychology 20 years on. *Cognitive Neuropsychology, 23*(1), 3–12.

Cipolotti, L., Butterworth, B., & Denes, G. (1991). A specific deficit for numbers in a case of dense acalculia. *Brain, 114*, 619–637.

Cohen, J. R., Asarnow, R. F., Sabb, F. W., Bilder, R. M., Bookheimer, S. Y., Knowlton, B. J., et al. (2011). Decoding continuous variables from neuroimaging data: Basic and clinical applications. *Frontiers in Neuroscience, 5*(75), 1–12.

Coltheart, M., Langdon, R., & Breen, N. (1997). Misidentification syndromes and cognitive neuropsychiatry. *Trends in Cognitive Sciences, 1*(5), 157–158.

Coltheart, M., Rastle, K., Perry, C., Langdon, R., & Ziegler, J. (2001). DRC: A dual route cascaded model of visual word recognition and reading aloud. *Spychological Review, 108*(1), 204–256.

Dawson, M. E., Schell, A. M., & Filion, D. L. (2000). The Electrodermal System. In J. T. Cacioppo, L. G. Tassinary, & G. G. Berntson (Eds.), *Handbook of Psychophysiology* (pp. 200–223). Cambridge: Cambridge University Press.

Downing, P. E., Jiang, Y., Shuman, M., & Kanwisher, N. (2001). A cortical area selective for visual processing of the human body. *Science, 293*, 2470–2473.

Ellis, H. D., & Hunt, R. R. (1993). *Foundamentals of cognitive psychology* (5th ed.). Dubuque, US: Wm. C. Brown Communications, Inc.

Ellis, H. D., & Lewis, M. B. (2001). Capgras delusion: A window on face recognition. *Trends in Cognitive Sciences, 5*(4), 149–156.

Epstein, R. A., & Kanwisher, N. (1998). A cortical representation of the local visual environment. *Nature, 392*, 598–601.

Fries, P., Nikolic, D., & Singer, W. (2007). The gamma cycle. *Trends in Neurosciences, 30*(7), 309–316.

Gross, J., Baillet, S., Barnes, G. R., Henson, R. N., Hillebrand, A., Jensen, O., et al. (2013). Good practice for conducting and reporting MEG research. *NeuroImage, 65*, 349–363.

Hagmann, P., Jonasson, L., Maeder, P., Thiran, J.-P., Wedeen, V. F., & Meuli, R. (2006). Understanding diffusion MR imaging techniques: From scalar diffusion-weighted imaging to diffusion tensor imaging and beyond. *RadioGraphics, 26*, S205–S223.

Hamalainen, M. S., & Ilmoniemi, R. J. (1994). Interpreting magnetic fields of the brain - minimum norm estimates. *Medical and Biological Engineering and Computing, 32*, 35–42.

Hari, R., Hämäläinen, M., Ilmoniemi, R., Kaukoranta, E., Reinikainen, K., & J., S. (1984). Responses of the primary auditory cortex to pitch changes in a sequence of tone pips: Neuromagnetic recordings in man. *Neuroscience Letters, 50*, 127–132.

Hearnshaw, L. S. (1987). *The shaping of modern psychology* (1st ed.). New York, US: Routlege & Kegan Paul Inc.

Iacono, W. G. (2011). Forensic "Lie Detection". *Journal of Forensic Psychology Practice, 1*(2), 75–86.

Kandel, E. R., Schwartz, J. H., Jessell, T. M., Siegelbaum, S. A., & Hudspeth, A. J., (2000). *Principles of neural science* (4th ed.). New York, US: McGraw-Hill.

Kempler, D. (2005). *Neurocognitive disorders in aging* (1st ed.). Thousand Oaks, California, US: Sage Publications, Inc.

Koutsouleris, N., Meisenzahl, E. M., Davatzikos, C., Bottlender, R., Frodl, T., Scheuerecker, J., et al. (2009). Use of neuroanatomical pattern classification to identify subjects in at-risk mental states of psychosis and predict disease transition. *Archives of General Psychiatry, 66*(7), 700–712.

Kutas, M., & Federmeier, K. D. (1998). Minding the body. *Psychophysiology, 35*, 135–150.

Legrenzi, P. (1997). *Manuale di psicologia generale* (2 ed.). Bologna, Italy: Il Mulino.

Logothetis, N. K. (2008). What we can do and what we cannot do with fMRI. *Nature, 453*, 869–878.

McCarthy, R. A., Evans, J. J., & Hodges, J. R. (1996). Topographic amnesia: Spatial memory disorder, perceptual dysfunction, or category specific semantic memory impairment? *Journal of Neurology, Neurosurgery and Psychiatry, 60*, 318–325.

Näätänen, R., Paavilainen, P., Rinne, T., & Alho, K. (2007). The mismatch negativity (MMN) in basic research of central auditory processing: A review. *Clinical Neurophysiology, 118*, 2544–2590.

Norman, K. A., Polyn, S. M., Detre, G. J., & Haxby, J. (2006). Beyond mind-reading: Multivoxel pattern analysis of fMRI data. *TRENDS in Cognitive Sciences, 10*(9), 424–430.

Peter, V., McArthur, G., & Thompson, W. F. (2010). Effect of deviance direction and calculation method on duration and frequency mismatch negativity (MMN). *Neuroscience Letters, 482*, 71–75.

Shallice, T. (1988). *From neuropsychology to mental structure*. Cambridge, England: Cambridge University Press.

Singh, K. D. (2006). Magnetoencephalography. In C. Senior, T. Russell, & M. Gazzaniga (Eds.), *Methods in mind* (pp. 291–326). Cambridge, Mass: MIT Press.

Tallon-Baudry, C. (2009). The roles of gamma-band oscillatory synchrony in human visual cognition. *Frontiers in Bioscience, 14*, 321–332.

Uhlhaas, P. J., & Singer, W. (2010). Abnormal neural oscillations and synchrony in schizophrenia. *Nature Reviews Neuroscience, 11*, 100–113.

Urgesi, C., Berlucchi, G., & Aglioti, S. M. (2004). Magnetic stimulation of extrastriate body area impairs visual processing of nonfacial body parts. *Current Biology, 14*, 2130–2134.

Chapter 2
Cognitive and Neural Aspects of Face Processing

Faces represent the stimuli we rely on the most for social interaction. They inform us about the identity, mood, gender, age, attractiveness, race and approachability of a person. This is remarkable if we think that all faces share the same composition of internal features (i.e., two eyes above the nose and a mouth) and 3D structure. Thus, faces are unique in terms of the richness of social signals they convey, and the reason why face perception has played a central role for social interaction in a wide range of species for millions of years. Given its importance, face processing has also become one of the most prominent areas of research in cognitive science of the last 50 years, and a large number of behavioural, neuropsychological and neuroimaging studies have significantly advanced our understanding of the developmental, cognitive and neural bases of face perception.

In this chapter we start our voyage towards the understanding of the features of human face recognition and we will learn about the most significant results on the cognitive and neural aspects of this fascinating topic. In particular, we start by analyzing this ability in normal subjects that do not show any sign of neurologic, psychiatric or neuropsychological disorder. Since it is important to know the normal features of a specific cognitive domain before we can start to comprehend the deficits, we will begin to address prosopagnosia, the disorder in face recognition, only in the next chapter.

Hopefully, after reading this chapter you would be able to answer important questions like: "Do faces represent a "special" category of stimuli for our visual system?", "How does face processing take place?", "Which is the neural underpinning of face processing?", "What is the speed of face processing?" and again, "Are face processing skills heritable?"

D. Rivolta, *Prosopagnosia*, Cognitive Systems Monographs 20,
DOI: 10.1007/978-3-642-40784-0_2, © Springer-Verlag Berlin Heidelberg 2014

2.1 Do Faces Represent a "Special" Category of Stimuli for Our Visual System?

2.1.1 Domain-Specific Hypothesis

In our environment we are surrounded by many different classes of visual stimuli such as cars, tables, chairs, churches, bottles, shoes, bodies and so on. The process of these objects (like all the other objects that you can have in mind) is mediated by *featural mechanisms*. This means that, put in simple words, we can recognize an object by combining all its features (tyres, windows, steering wheel) together (it is a car!).

Can we use the same featural mechanisms to process faces? In other words, can we recognize a face by putting all its features together? The answer is yes; we can do it. We can focus on the eyes or the nose of a person in isolation and recognize their "owner". However the evolution, over thousands of years, also equipped us with *holistic processing*.[1] Holistic processing enables us to perceive a face as a gestalt (a whole), which is more than the sum of the individual components. There is mounting empirical evidence showing the existence on holistic processing and researchers are starting to understand that holistic mechanisms are important for typical face processing. Below I will describe some of the experiments that (indirectly and directly) demonstrated holistic processing for faces. In particular, I will describe the face-specific effects that these experiments have shown.

Yin (1969) described for the first time what has been referred as the *face-inversion effect*.[2] This effect indicates that, in experimental environments, if people have to learn and remember faces they have never seen before (i.e., unfamiliar faces) they would be 20–25 % better at doing it when faces are shown upright than upside down (see Fig. 2.1).

Of course, you may think, that this is true even for objects. In fact we do not live in an upside-down world and inversion would affect the recognition of objects as well. This is true; the inversion effect occurs even for objects, but it is *much smaller* (up to 8 %) than the one shown for faces (see a very informative review on the topic in McKone et al. 2009). Yin suggested that this disproportionate inversion effect for faces is due to the fact that extracting the correct relationship between the face parts (holistic processing) was particularly important to face recognition and that extracting this information from inverted faces was difficult. In summary, according to this first key experiment, holistic processing does not "work" for inverted faces and it is not critically involved in object recognition.

---

[1] Over the past 20 years different authors have used different names such as configural, second-order relations or global to refer to what I define here as holistic processing. The theoretical reasons behind this go beyond the aim of the book.

[2] Also known as the *disproportionate inversion effect for faces*.

Fig. 2.1 It is difficult to recognize faces when they are shown *upside-down*. This picture represented here shows how a grotesque alteration of a face (try to turn the book *upside-down*) almost gets unnoticed on an inverted face (picture courtesy of Dr Rachel Robbins)

Tanaka and Farah (1993) provided more direct evidence in favor of holistic processing. They prepared a task (see Fig. 2.2) where people had to learn different identities such as "Tim" and were subsequently asked to recognize some of his features (e.g., the nose) in a forced choice task where Tim's nose had to be distinguished from another person's nose (e.g., Bob). Sometimes the two noses were shown in isolation and subjects had to indicate which one was Tim's nose (Fig. 2.2, top); sometimes the noses were shown within a face (Tim's nose in Tim's face and Bob's nose in Tim's face, see Fig. 2.2, bottom). The task was to indicate whether Tim's nose was on the left or on the right of the computer screen. Results indicated that the identification is better when features are shown within the face than when they are in isolation. This is called the *part-whole effect*. Importantly, this effect disappears (or is much smaller) in upside-down faces and with objects. Once again this effect suggests that upright faces only can benefit from holistic processing, since the face contour dramatically cues the correct recognition of a feature (nose) in it (McKone et al. 2009).

Another effect psychologists are very familiar with is the *composite-face effect*. This effect can be demonstrated in tasks where participants have to indicate whether, for example, the top-part of two sequentially presented unknown face stimuli is the same or different. Stimuli are shown in two different conditions: aligned or misaligned (see Fig. 2.3).

Results from different studies indicate that people are faster and more accurate when the halves are misaligned. This is because aligned faces (even when made up of two different identities) automatically "glue-up" to form a new configuration

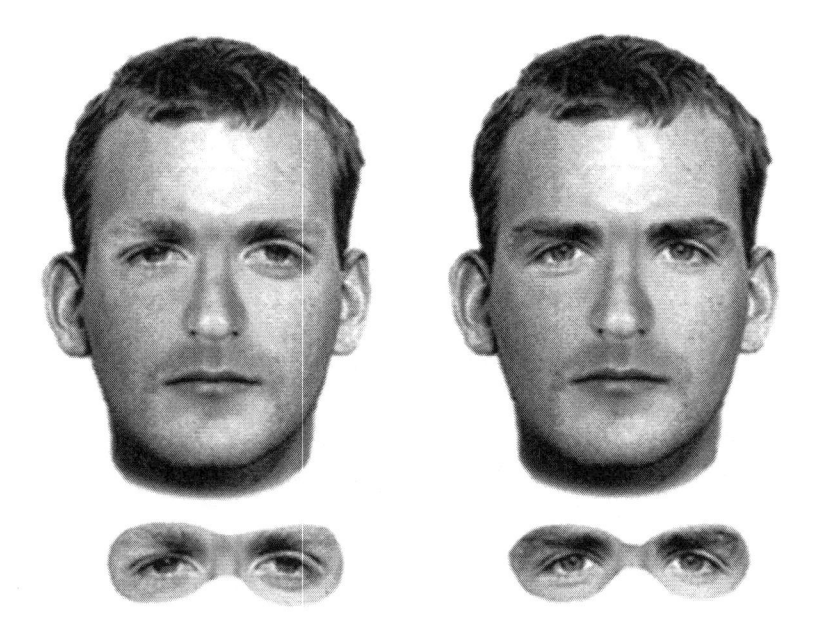

Fig. 2.2 The part-whole effect demonstrates that the identification of a particular feature (e.g., eyes) is facilitated when this is presented within the face configuration (*top*) then when in isolation (*bottom*). Figure obtained with permission from Palermo and Rhodes (2002)

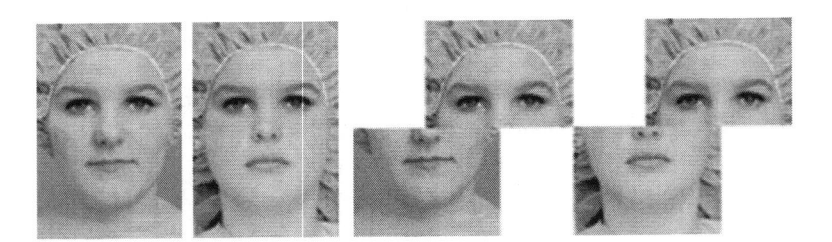

Fig. 2.3 The composite-face effect. Judging whether the *top halves* of two pictures depict the same person (as in this figure) is harder when the two halves are aligned (forming a new identity) than when they are misaligned (these face stimuli were kindly provided by Prof. Daphne Maurer from McMaster University, Canada)

(a new identity). It is then extremely difficult to process one half of the face without being influenced by the perception of the other half of the face, which makes the task more difficult. Once more, this effect is absent for other objects and for upside-down faces, strengthening the dedicated role played by holistic mechanisms in upright face perception (Robbins and McKone 2007; Young et al. 1987).

The last approach I wish to describe for the assessment of holistic processing uses artificially modify faces to change the spacing between features (for example the distance between the eyes) or the features themselves. In a very well-known

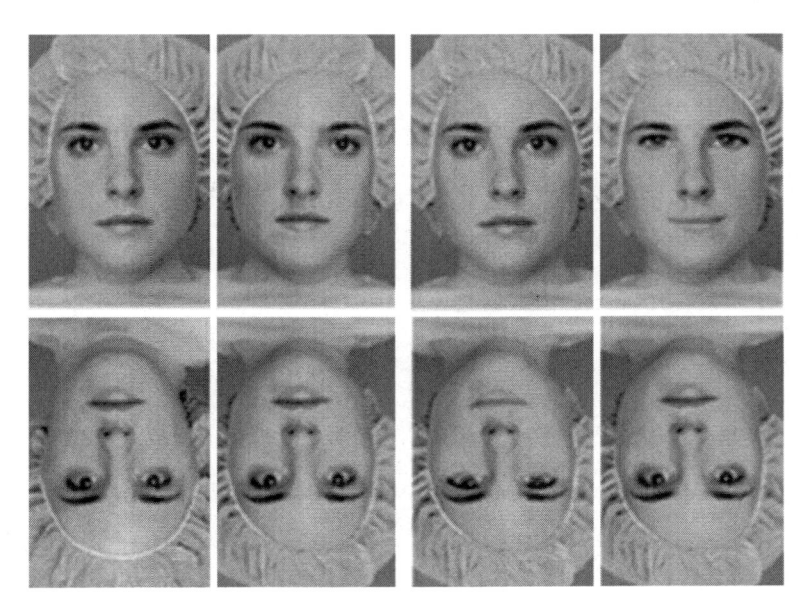

Fig. 2.4 Examples of stimuli used in the Jane task. It is harder to detect the differences in the "spacing condition" (*top* and *bottom left*) than in the "features" condition (*top* and *bottom right*) when faces are upside down (these face stimuli were kindly provided by Prof. Daphne Maurer from McMaster University, Ontario, Canada)

task called the "Jane task" (Le Grand et al. 2004; Mondloch et al. 2002) participants had to determine whether two sequentially shown faces where the same or different, when the difference was either in the spacing of the features or in the features themselves (see Fig. 2.4).[3]

The same task is then given with upside-down stimuli. Results showed that performance decreases, overall, for upside-down stimuli, but that this inversion affects the detection of changes to spacing more than the detection of changes to features. Since, as we should know by now, face inversion is supposed to affect holistic processing, this result on the Jane task is in agreement with the idea that holistic mechanisms process the spacing between face parts and not the parts themselves.[4]

What we have learned so far is that faces, inverted faces and objects are processed by *featural mechanisms*. However, the perception of upright faces is mediated by *holistic mechanisms* more than any other type of stimuli. The effects reported above strongly support this claim and, as such, support what is known

---

[3] In the original Jane task there is also a third condition, called the "contour condition", that for clarity reasons I do not report here. The interested reader is invited to refer to the original articles.

[4] Some researchers believe that holistic processing involves even the features of faces, but it is beyond the aim of the book to address those theoretical issues. See McKone and Yovel (2009) for a detailed description of the issue.

as the *domain-specificity hypothesis*. The domain-specificity hypothesis states that faces represent a "special" category of stimuli that are processed by holistic mechanisms (McKone et al. 2006). Without these mechanisms our recognition would be problematic, as we will see in the next chapter. You may think that the experiments I showed you above lack of *ecological validity* that is, our visual system is never exposed to those stimuli in everyday life. This is in part true; how often would you see a misaligned composite face at the pub? However, in order to study our cognition we need somehow to decompose our complex processes (e.g., visual cognition) into smaller and investigable units. This is the reason why each of the tasks I presented above focuses only on one specific aspect of face processing. In addition, this process of division in subtasks enables us to have a rigid control over the variables under investigation.

2.1.2 The Expertise Hypothesis

There are researchers that do not support the domain-specificity hypothesis. They support the *expertise hypothesis*; which states that expertise plays a critical role in developing holistic mechanisms for faces. In general, the basic difference between face and object perception is the "depth" of processing: when we see a face we can identify it, whereas when we see a table we usually do not. According to this view, experts in a particular field such as dog experts (e.g., people who can individually identify many different individual Golden Retrievers) or car experts (i.e., people that can identify, in a glance, many different makes of cars) should show holistic processing not only for faces but even for their category of expertise. Even though some early studies supported the expertise hypothesis, more recent and better controlled experiments failed to support the expertise hypothesis, suggesting that only upright faces rely on holistic processing (McKone et al. 2006, 2009).

Let me give you some examples of the evidence discarding the expertise hypothesis (experiments reported in Robbins and McKone 2007). In these experiments the authors tested dog-experts (people with, on average around 23 years of experience as dog judges, breeders or trainers) and novices, that is, non-experts in dog recognition. In theory, both experimental groups are expert in face perception, but only dog-experts are also expert in dog perception. Accordingly, following the expertise hypothesis, this manipulation should demonstrate holistic effects not only in face processing but also in dog perception for dog-experts only. In one of these experiments dog experts and novices had to memorize (upright) dogs and faces. After an interval of few minutes where they had to complete a different task, the memorized stimuli were shown on the screen with a distractor (i.e., a stimulus that was not shown during the learning phase); this means that, for each trial, each subject hat to decide whether the learned stimulus was on the left or on the right of the computer screen. The procedure was repeated when stimuli were shown upside-down (rotated 180°). Results demonstrated that, as expected, novices showed a face inversion effect; that is, the inversion affected more the memory

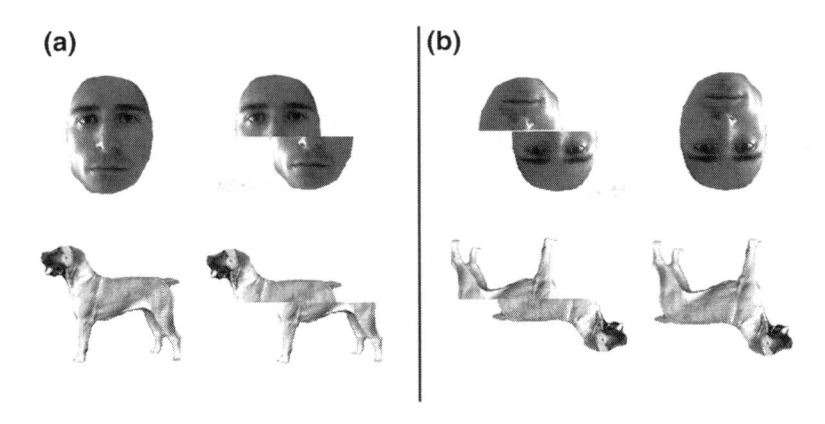

Fig. 2.5 Facsimile of stimuli adopted in the composite-face task. Faces and dogs were shown in the aligned and misaligned condition both (**a**) upright and (**b**) upside-down (These stimuli were not the actual one adopted in the original experiments. I created them for representational purposes only)

performance for faces than for dogs. Interestingly, despite their great experience with dogs, even dog-experts showed bigger face inversion than dog inversion effects. This strongly supports the domain-specificity hypothesis (i.e., faces are special and processed by holistic processing) and not the expertise hypothesis (i.e., all stimuli with which subjects had strong experience can be processed like faces, using holistic processing).

Another experiment addressed whether the composite face effect could be demonstrated with stimuli of expertise (i.e., dogs) in dogs-experts. The procedure is similar to the composite-face task I presented above. As in classical composite-face tasks, the faces were shown aligned and misaligned, in both upright and inverted conditions. The same happened for dogs, they were presented aligned and misaligned stimuli, both upright and upside down. Results demonstrated that novices showed, as expected, the classical result: they were more accurate when the halves were misaligned then when they were aligned, that is, they showed a composite-face effect (see Fig. 2.5 for a description of the results). Similarly to many other studies, results showed that this effect disappeared when stimuli were shown upside-down. Note that in both orientations they did not show any composite effect for dogs; this was expected since novices had no experience in dog recognition whatsoever. Will dog experts show the composite effect for dogs, the class of stimuli they have a lot of expertise with? Results were negative; dog-experts showed a composite effect for faces only and not for dogs. Similar to novices, they did not show any composite effect for inverted stimuli (Fig. 2.6).

Overall these two experiments, along with many others, demonstrate that holistic mechanisms are features of upright faces processing only. This is not because we acquire expertise with faces during our lives, but because, as we will see in the

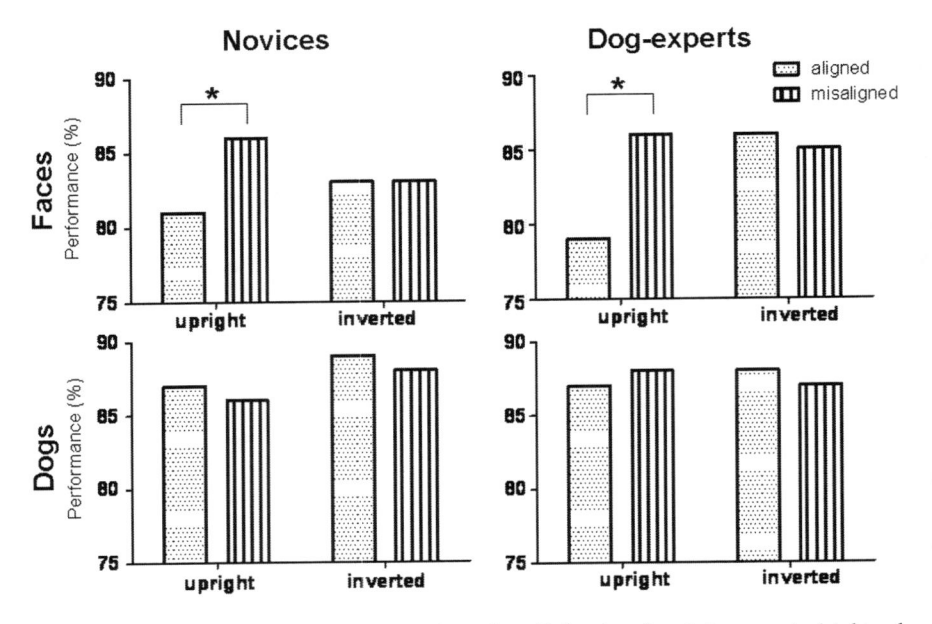

Fig. 2.6 Results on the composite task. Both novices (*left column*) and dog-experts (*right column*) show a composite effect for faces (the *asterisk* indicates a statistically significant difference between the two conditions). No group shows a composite effect for dogs, suggesting that expertise with a category of stimuli does not develop holistic mechanisms for that category (This figure is not taken from the original work of Robbins & McKone (2007), but it only wishes to represent, with fictitious data, the main effects described in the original work)

next paragraphs, we might be born with this ability. Thus, the answer to the question: "Do faces rely on "special" cognitive processes?" is *yes*. The processing of upright faces only relies on dedicated cognitive mechanisms. This makes faces a *special* stimuli for the human visual system.[5]

2.2 How Does Face Recognition Take Place?

It takes us a small fraction of a second to recognize a face once we perceive it. Even though we experience every day the speed of our face processing system, it is known that the recognition of a familiar face does not occur all at once, but involves different cognitive steps. In 1986 two British researchers (Vicki Bruce and Andy Young) described those steps in an influential cognitive model of face processing known as the "Bruce and Young model of face recognition" (Fig. 2.7). The model was developed to accommodate a range of empirical observations from

---

[5] The reader interested in the theoretical debate between the competing domain-specific and expertize hypothesis is invited to read works from Gauthier and colleagues.

Fig. 2.7 A simplified version of the Bruce and Young (1986) model of face recognition. This model depicts the stages between the perception of a face and its recognition (*FRUs* face recognition units; *PINs* person identity nodes; see text for a complete description of the model)

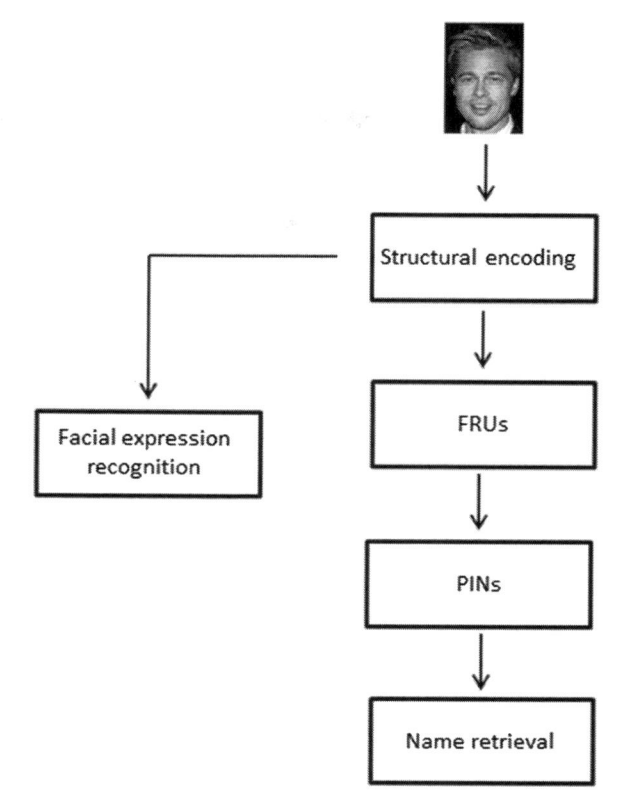

normal subjects and neuropsychological patients, and proposed an organization of the cognitive face processing system that is hierarchical and branching.

The model proposed that face recognition occurs via an initial stage of *Structural Encoding* of a face's appearance, where both holistic and featural processing are supposed to occur (Schmalzl 2007). This is followed by firing of *Face Recognition Units* (FRUs) that respond to particular features and configurations of specific familiar faces, access to relevant semantic information within the *Person Identity Nodes* (PINs), and finally name retrieval. In addition, it proposed that this process of person identification occurs independently from the identification of facial expressions and lip reading. Bruce and Young postulated that the representations of identity and those of the more changeable aspects of a face (i.e., emotion expression) must be (at least to some extent) independent from one another, or else a change in expression or a speech-related movement of the mouth could be misinterpreted as a change of identity.

With an example, we can see that Brad Pitt's face is processed within the structural encoder that codes for the structure of the face such as having two eyes above the nose above the mouth, and the distance between them (we will describe in more detail this in the next paragraph). This information is fed to the FRUs that "fire" because the face is familiar and then to the PINs where biographical information such as that 'this person was married to Jennifer Aniston, is currently (January 2013) married to Angelina Jolie, starred in Troy, Fight Club, The curious case of Benjamin Button' and so on. Finally we can retrieve the name: "Brad Pitt!" According to the model, the understanding of whether the face of Brad Pitt depicted in the photo looks happy, angry, sad or disgusted is mediated by a different system. In Chap. 3 we will see how this model can account for problems in face perception.

2.3   What is the Neural Underpinning of Face Processing?

When you look at an object from the outside world, the light reflects from it and bounces into our eyes, where a structure called the retina decodes this information and passes it to the brain. As discussed in Chap. 1, the occipital lobe at the back of our brain is the main anatomical region that receives this information. From the occipital lobe, information is diverted to other brain regions such as the temporal lobe and the parietal lobe (Milner and Goodale 2006) (Fig. 2.8).

We have known for a few decades that the temporal lobe is crucial for our ability to recognize faces, objects and places. Let's focus now on faces. The first question here is: "How can we know about the crucial role played by the temporal lobe in face recognition?" Single unit recordings (the technique that consists of directly placing electrodes in the brain, through the skull, and measuring the firing rates[6] of neurons) in anesthetized monkeys paved the way for the current knowledge we have in face recognition. In 1969 Gross and colleagues recorded the activity of neurons in the inferior part of the macaque temporal cortex. They found that there were groups (patches) of neurons that responded virtually *only* when the animal was looking at faces and not when it was looking at fruits, hands or tools. Some of these neurons responded to front view faces, other to profile faces, other to face parts (i.e., eyes). There were even single neurons that responded to some specific identities (people that monkeys were familiar with such as the researchers working with them). These results collected over 50 years ago strongly suggested that the face recognition system, at least in monkeys, has developed some modules that process faces only, and they provided the first anatomical substrate that underlies the special status that faces play for social functioning (see a review on the topic in Gross 2008).

Only after many years, with the advent of modern functional neuroimaging techniques such as positron emission tomography[7] (Haxby et al. 1994) and fMRI, researchers have attempted to localize face recognition processing within the

---

[6] Firing rates are typically measured as the number of action potentials (*spikes*) a neuron fires in 1 s.

[7] PET belongs to the class of invasive neuroimaging techniques. This technique enables to see brain activity only after the intravenous injection of a radioactive substance. PET represented one of the most adopted techniques for the visualization of brain activity in vivo before fMRI was invented.

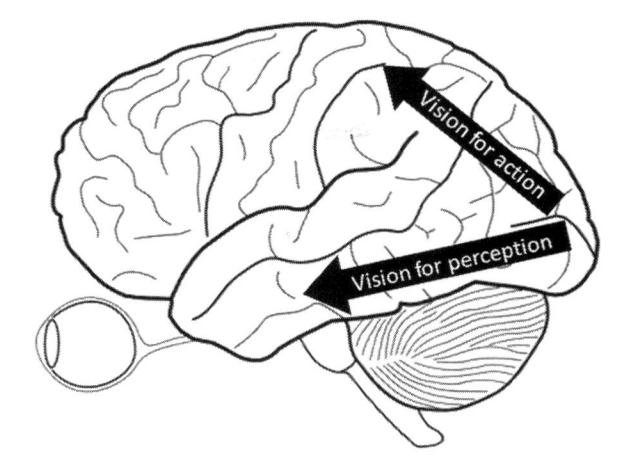

Fig. 2.8 Lateral view of the left hemisphere. The *two arrows* indicate the two visual systems: the one we use for perception (*vision for perception*) and the one we for action (*vision for action*). The first, also known as occipito-temporal route, mediates objects and face recognition; the second one, also known as occipito-parietal route, mediates visually guided movements, such as reaching a glass

human brain. One of the most influential findings was the discovery made by Kanwisher and colleagues (1997). By using fMRI the authors identified and localized the so called Fusiform Face Area (FFA) within the human temporal lobe. In this well-known experiment people were shown faces, cars, hands and other objects, while their neural (BOLD) activity was recorded. Results showed that, similarly to findings in monkeys, there was a region (the FFA) within the human temporal lobe that showed a response for faces that was at least twice as strong as for other objects. This seminal research was subsequently confirmed in many other studies leading to other research questions such as whether the FFA represents the region where holistic processing, the face-dedicated processing, takes place. It is now believed that FFA represents a crucial region for holistic processing, since the FFA responds stronger for upright face processing than inverted face processing[8] and it mediates the composite face effect (remember that face inversion and the composite face effect are believed to demonstrate holistic processing) (Liu et al. 2009; Schiltz and Rossion 2006; Yovel and Kanwisher 2005). In addition to holistic processing, however, it has been shown that the activity of FFA is also involved in face identification and recognition (Rotshtein et al. 2005). It should be noted however that further research failed to confirm the role of FFA in face identification (Kriegeskorte et al. 2007) and further studies are needed to clarify the issue.

Over the last 20 years, numerous cortical face-sensitive brain regions have been discovered in humans. Each of them seems to represent the neural correlate of different behavioural phenomena. The occipital face area (OFA, Gauthier et al. 2000) in the occipital lobe seems to respond mainly to face features, the superior temporal sulcus (STS) to changeable aspects of the face such as facial expression, and the anterior temporal face patch (ATFP) of the anterior temporal lobe to face identities (Gobbini and Haxby 2007; Haxby et al. 2000; Kriegeskorte et al. 2007) (Fig. 2.9).

---

[8] Face inversion increased the activity of object selective regions, further suggesting that inverted faces are processed using mechanisms that are common to object processing.

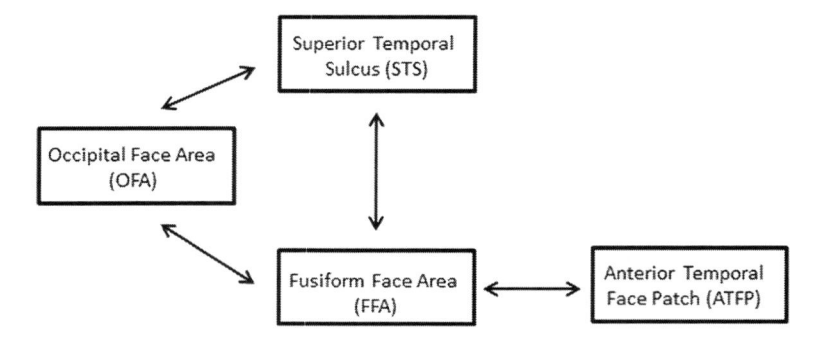

Fig. 2.9 Simplification of the Haxby et al. (2000) neural model of face processing. This model schematically represents the main human neural substrates for normal face processing. The OFA may be the first "gate" of the system, where a stimulus is judged to be a face and where its features are processed. This information bifurcates to the STS for the process of emotions (and other changeable aspects of faces such as eye gaze) and to the FFA, where holistic processing (and maybe identification) is carried on. The identification of faces occurs in (but not only) the ATFP

Causal evidence for the crucial role played by OFA in the processing of face features comes from the use of TMS, which, as seen in the previous chapter, temporarily "deactivates" a specific brain region and monitors the consequent behaviour. Pitcher and colleagues (2007) presented participants with faces and houses that changed in the shape of their features (e.g., different eyes, different windows) or in the spacing of their features (i.e., the distance between features changes, but the features themselves remained the same). As the reader may remember this manipulation is similar to the one described for the Jane task above. The idea behind the manipulation of this study is that the spacing-change detection reflects holistic process, whereas feature-change detection reflects features processing. Participants had to determine whether two sequentially shown pictures of faces or houses depicted the same stimulus or not (that is, if there was a spacing change or a feature change or no change). Results demonstrated that TMS delivered within 60–100 ms post-stimulus onset on OFA (previously determined for each subject individually using fMRI) disrupted the face-performance on the feature part of the task, and not the spacing task. The specificity of this disruption for face processing was demonstrated by showing the total absence of TMS stimulation consequences on house discrimination. In summary, this study showed the critical role played by OFA in the processing of face features; so OFA does not seem to represent the locus of holistic face processing.

The importance of STS for changeable aspects of a face was given by Narumoto and colleagues (2001). The authors, amongst other conditions, asked participants to place their attention to the expression or the identity of faces they were shown. Results obtained with fMRI indicated that the (right) STS was more involved in the coding of facial expression than identity, suggesting the involvement of STS in changeable aspects of the face (and not in identity discrimination). The role of the ATFP for face identification has recently been given, using fMRI, by Kriegeskorte and colleagues (2007). Authors demonstrated that the pattern of fMRI activity within this region, but not FFA, could discriminate between two different identities participants had to learn. Overall, the differential role played by these face areas in humans has until now not definitely been clarified and future research will address the issue (in Chap. 3 we will see how these face regions are involved in disorders of face recognition). What seems clear from previous research in cognitive science is that the right hemisphere of the human brain is dominant for face processing. Dominant means that great part of the work in face recognition is done by the right hemisphere.

The existence of face specific clusters of neurons in the human brain has not only been determined by fMRI, but it has been very often confirmed even by (invasive) single unit recordings in patients with drug-resistant epilepsy. These patients cannot find relief from their condition with the use of anti-epileptic drugs and they sometimes need surgery aimed at the removal of the neural tissue generating the epileptic seizures (*focus epilepticus*). In order to remove the correct region, some invasive recordings are necessary and researchers have the unique opportunity to run exceptional experiments aimed at clarifying the relation between the mind and the brain (Duchowny 1989).

In 2005 it was discovered the existence of neurons in the human (medial) temporal lobe that respond preferentially to some specific identities. For example authors described a neuron that fired with Jennifer Aniston's (a famous American actress) pictures only. Different pictures of the actress taken many years apart with different hairstyle and different orientation of the face elicited the response of these neurons. Importantly those neurons did not respond to faces of other celebrities. Other groups of neurons responded only to other famous people such as Michael Jordan (a famous Basketball player) or Bill Clinton (a former US president). Furthermore, recent evidence in two cases of people with drug-resistant epilepsy demonstrated that the electrical stimulation of face-sensitive regions within the fusiform gyrus (Parvizi et al. 2012) and within the lateral occipital gyrus (Jonas et al. 2012) completely distorted face recognition, thus leading to "transient prosopagnosia".

In summary, there is a convergence from animal and human studies that the face recognition system relies on dedicated neural populations. This, once again, suggests that faces are "special"; even at the neural level. What happens at the behavioural level when the brain regions involved in face processing get damaged will be the topic of the next chapter.

2.4   What is the Speed of Face Processing?

2.4.1   Neurophysiological Investigations

As we have just learned, there exist regions within the human brain that are specifically "tuned" for face processing. However, we still do not know how long it takes for these regions to compute the processes. Are these regions fast at processing? How fast?

As explained in Chap. 1, EEG and MEG (but also TMS) are the best available non-invasive techniques to adopt for answering this question. Since the late 80's it is well known that the existence of an electrophysiological response for faces that has been recorded both with intracranial invasive methods in people with drug-resistant epilepsy (Allison et al. 1994; Allison et al. 1999) and with surface EEG (Bentin et al. 1996; Jeffreys 1989). When people are shown faces, their EEG activity shows a negative deflection occurring at around 170 ms (ms) post stimulus onset; this is known as the N170. The N170 is detectable from occipito-temporal surface electrodes and consistently shows bigger amplitude for faces than other categories of visual stimuli. It is believed that the N170 is mainly generated by the activity of two cortical regions such as the occipital lobe (OFA) and the temporal lobe (FFA), where there are neurons "tuned" for faces (Deffke et al. 2007; Itier et al. 2007; Linkenkaer-Hansen et al. 1998). Since these brain regions have face-sensitive neurons, it is believed that the synchronous firing of thousands of those neurons can give rise to a potential that is big enough to be seen on the scalp (N170). As expected, the face-sensitive activity at around 170 ms post stimulus onset has been detected even with MEG and it is named M170 (Liu et al. 2000). Albeit, as we just said, the generators of the N/M170 are two, it is still largely unexplored whether each of these two regions can generate a distinct N/M17. In other words: "Can the two sources of the N/M170 code for a different kind of face processing?" I will answer this question in the next chapter when I describe congenital prosopagnosia. For now we focus on a classical finding. The peak of the N/M170 is delayed of 10–13 ms when faces are presented upside-down. This inversion effect of the N/M170 suggests that holistic processing occurs at around 170 ms post-stimulus onset (Bentin, et al. 1996; Rossion et al. 2000).

The N/M170 represents by far the most explored face-sensitive electrophysiological component. However we know that there is a face-sensitive component that peaks earlier than the M170; this component peaks at around 100 ms post stimulus onset and since it has been firstly investigated with MEG, it is called the M100 (Liu et al. 2002). This component is generated from the occipital lobe and supposedly codes for aspects of face processing distinct to the M170. One hypothesis is that M100 reflects the detection that a face is present in the visual field, whereas the N/M170 enables the identification of it (Liu et al. 2002). Other lines of evidence suggest that the M100 codes for face features, whereas the N/M170 is sensitive to holistic processing (Pitcher et al. 2011), albeit some evidence showed that the N/M170 codes for both features and holistic processing (Harris and Nakayama 2008). One of the main reasons for these discrepancies in results may rise from the different technique (EEG versus MEG), experimental design, methodology and data processing. One recent study shed further controversy on the topic by indicating that even the M100 can be sensitive to face familiarity (Rivolta et al. 2012). Future studies will hopefully clarify the issue.

2.4.2 Behavioural Investigations

This impressive speed our face recognition system shows for face processing further supports neuroimaging and behavioural data claiming that faces have a special status in our face recognition system. Since we already described the most important neuroimaging findings in face processing research, let's focus our attention on some behavioural experiments that attempted to understand how long it takes to determine that there is a face in a visual scene and how long it takes to identify a familiar face. By using a Rapid Visual Serial Presentation (RVSP) paradigm, characterized by the rapid and sequential presentation of visual stimuli, it has been shown that object categorization (i.e., deciding whether a visual stimulus is a face, an animal, or an object) occurs just as rapidly as the mere detection of an object within a visual field. Since the effect occurred even when stimuli were shown for only 17 ms on the screen, this result strongly suggests that object detection occurs as quickly as its categorization, thus indicating that stimulus detection and categorization may occur in parallel (Grill-Spector and Kanwisher 2005; Purcell and Stewart 1988). This result is in line with the common experience that as soon as we see something we can say that it is a face (or an object). Interestingly, in agreement with MEG results described above, it has even been demonstrated that face identification can occur in around 100 ms (Tanaka 2001). Further evidence supporting the exceptional speed of face processing comes from investigations that monitor the eye movements. Using a specific device, called the *eye-tracker*, it is possible to monitor the speed and the features of eye movements. In other words, it is possible to see where and for how long people focus their sight on visual stimuli presented on a computer screen. It has been recently shown that when people have to make eye movements towards target stimuli such as faces, animals and vehicles, they are on average more accurate and much faster when they have to do it for faces than other categories of visual stimuli. In addition, the minimum saccadic reaction time towards faces occurs in 110 ms, faster than for animals (120 ms) and vehicles (140 ms) (Crouzet et al. 2010).

2.5 Are We Born with "Face-Specific" Cognitive and Neural Mechanisms?

As described at the beginning of this chapter, over the past decades there has been ongoing debate about whether face specific cognitive mechanisms, that is holistic processing, is acquired or it is present since birth. One way to find the solution to this issue is by investigating face processing in very young kids; infants in particular. For practical reasons the methodologies adopted for research in infants are very different from the ones adopted for adults (would you expect a 4 day-old infant to perform the composite-face task?). Typically, *looking time* (i.e., the infant spends more time fixating something of interest than something less interesting) measures are adopted as an index of preference.

Research accumulated over the last 10 years strongly suggests that humans are equipped with face-specific cognitive mechanisms *from birth*. Until recently however, it was believed that children need around 10 years of experience with faces to show the face-specific experimental effects described above in adults (Carey et al. 1980). This, which was initially taken as strong support for the expertise hypothesis (Diamond and Carey 1986), has subsequently been demonstrated to be wrong (see McKone et al. 2012; McKone et al. 2009 for an excellent and in detail description of the issue).

2.5.1 Behavioural Studies in Infants

Behavioural studies have indicated that newborns show a preference for tracking face-like configurations compared to other types of visual arrays (Fig. 2.10). Three-month-olds (and even 3-days-old!) can recognize the identity of novel individuals with similar looking faces (presented without hair and across view changes), suggesting that a face representation tuned to upright faces and able to support individual level representations is present at birth (Pascalis et al. 1998; Slater et al. 2000; Turati et al. 2008). Newborns less than a week old prefer attractive over unattractive (as rated by adults) faces when stimuli are upright, but not when inverted (Slater et al. 2000).

In addition, children as young as 4 years of age can show an inversion effect, a part whole effect, a composite effect, as well as general sensitivity to spacing of face features (for example as tested on the Jane task) (Cohen and Cashon 2001; Hayden et al. 2007; McKone et al. 2009). In addition, these effects not only are presents in young kids, but (by around 5 years of age) also show similar magnitude of the effects seen in adults. That is, the effect is not present, but has the same size as the one observed in adults (McKone et al. 2012). In summary, so far, all behavioural findings of adults' face processing are present with similar strength in infants. Thus, developmental studies strongly suggest that years of experience and extensive practice with face individuation cannot be the only factor that accounts for the observation of face specific experimental effects (i.e., the expertise hypothesis cannot be supported).

Furthermore, all infants (and even monkeys) show *perceptual narrowing*. It has been demonstrated that infants have, at birth, the capacity to represent all faces (faces from all races and even monkey faces!). This ability narrows down during time and infants become specialized for types of face that they are exposed to frequently in their native environment. Results showed that 6-month-old infants could discriminate both humans and monkey faces, while 9-months-old and adults could only discriminate human faces (Pascalis et al. 2002). Perceptual narrowing for faces explains why we tend to be better at perceiving, memorizing and identifying people from our own race than other races. This effect is called the

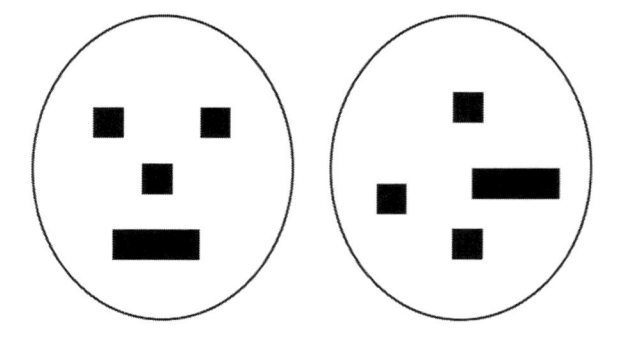

Fig. 2.10 Research with newborns has adopted stimuli like these. On the *left* there is a face-like configuration, whereas on the *right* there is a configuration that, despite showing the same elements, does not resemble a face. Newborns spend more time looking at the stimulus on the *left* than the one on the *right*

other-race effect (Feingold 1914). If you are European and grew up in Europe, when travelling to Asia for the first time you may find it more difficult to differentiate between Asian people than you do for people in your own country. The same is true, of course, for Asians that come to Europe for the first time. Perceptual narrowing for faces is similar to what happens in the domain of language (Kuhl et al. 2003), where newborns can discriminate phoneme boundaries from all possible languages (i.e., we can potentially learn all languages at birth), but they lose this ability with time. At first the perceptual narrowing for faces seems a negative aspect of development, since we are losing something that might be useful; however, this specification toward a particular race enhances our ability to discriminate people within this race.

The last aspect I wish to underline is the idea of the existence of a *critical period* (Sengpiel 2007). Experimental evidence suggests that face processing is characterized by a critical period, requiring adequate environmental input (i.e., normal face perception) before the face-system is used for other purposes. What happens if we do not show faces to infants for a few months (or years)? Of course we are not so cruel as to do this to our kids, but children with *congenital cataracts* in the eye/s do not receive proper visual inputs until the cataracts get removed (usually around 1 or 2 years old). Le Grand and colleagues (2004) demonstrated that even many years after the removal of the cataracts (even 12 years as indicated in Ostrovsky et al. 2006), years in which they had normal exposure to faces, people showed impaired holistic processing (e.g., impaired performance on the composite-face task and on the Jane task).

In summary, behavioural data reviewed above demonstrating adult-like face abilities present at birth, perceptual narrowing and critical periods, are all consistent with a genetically determined *innate* contribution to infant face recognition.

2.5.2 *Neuroimaging Studies in Infants*

A second line of evidence supporting the innate origin of our face recognition abilities comes from neuroimaging investigations. Many studies have shown, as described above, that a region in the temporal cortex, the FFA, responds more robustly to faces than to other types of visual stimuli. Similarly, studies using M/EEG indicated that brain responses occurring approximately 170 ms after stimulus onset typically show greater amplitude for faces than for other types of objects. When tested using fMRI, around 80–85 % of children between 5 and 10 years of age showed an FFA. In developmental studies, the size of (right) FFA correlated with behavioural performance on face memory task but not with object memory. In addition, ERPs studies in young children reported a face-specific N170 that showed even a face-inversion effect (Scherf et al. 2007).

Altogether the results reviewed so far demonstrate that both behavioural and neuroimaging findings reported in adults can be found in the youngest age group tested. However, even if the size of behavioural effects and the N170 seem to be comparable with adults, the magnitude of FFA activity is less clear and will require further research. Overall, research seems to support the claim that humans are born with innate face-specific cognitive and neural mechanisms.

2.6 Are Face Processing Skills Heritable?

It is well known in psychology that people show variability in their performance on different cognitive tasks. This variation represents the base of *individual differences*. Some of us are very good at calculations other are less good; some people have very good memory skills, others do not, and this also goes for reading abilities, athletic performance, musical skills, etc. Are we variable in face recognition abilities or everyone has the same skills? Similar to other abilities, many studies have underlined the existence of strong individual differences in face processing (Bowles et al. 2009). A key question regarding the understanding of whether this variability in face processing skills is "genetic", heritable, or whether the environment (i.e., the exposure to faces, social status of the family, country of origin and so on) plays a critical role in shaping our skills.

It is known that general intelligence is highly heritable in humans. Can, however, a specific ability such as face recognition show an high level of heritability? In other words, can someone inherit high intelligence, but poor face recognition skills from the parents? (McKone and Palermo 2010). Twin studies constitute an interesting methodological approach to the issue. Twins can be monozygotic or dizygotic. Monozygotic twins share the 100 % of their genes, whereas dizygotic twins share around 50 %. This means that, since twins usually share the same familiar environment, a difference in the correlation between specific measures (i.e., face memory) in the two groups must be attributed to genetic and not environmental factors.

Wilmer and colleagues (2010) looked at face memory skills in 164 monozygotic and 125 same gender dizygotic twins. Results showed that there is a correlation between performances of monozygotic, but not for dizygotic twins; this means that if twin A of a monozygotic twin couple performed very well, twin B tended to do the same. On the other side, if twin A of a dizygotic twin couple performed very well, performance of twin B was not necessarily also good. Since monozygotic and dizygotic twins share the same environment, this difference posits for genetic factors in individual variations in face processing. The question of which genes are involved however remains unanswered. Importantly, both monozygotic and dizygotic twins were not correlated in their performance on a memory task that did not tap into face processing such as abstract art memory or a paired-associates memory test, indicating that face processing skills *only* are heritable and they do not depend on general attention and/or memory functioning. Overall these data add to previous results strongly indicating special cognitive and neural mechanisms for face processing. In the next chapter we will further discuss the role of familiarity in face processing skills by discussing cases of congenital prosopagnosia.

2.7 Conclusions

In this chapter we have learned that upright face processing only is mediated by specific cognitive (i.e., holistic mechanisms) and neural (e.g., OFA and FFA) mechanisms. In addition, face processing is mandatory, occurs very quickly and it is mediated by face-sensitive physiological mechanisms (e.g., M100, M170). All these features seem to be present from birth and not acquired (although they may be improved) with experience. Overall, the evidence reviewed in this chapter strongly indicates that faces represent *special* stimuli for our visual system. In other words, faces seem to represent the category of visual stimuli that engage the fastest and most dedicated cognitive and neural processing.

In the next chapter we will learn that this special and precious ability most of us share can fail, causing serious and embarrassing problems in face recognition. Thus we will talk about people that have lost their ability to recognize faces after brain injuries (*acquired prosopagnosics*) and about people that have never developed the typical ability to recognize face (*congenital prosopagnosics*). We can anticipate that since prosopagnosics typically have specific problems in face processing while their object processing is spared (or much less impaired than face processing), the existence of prosopagnosia further supports the special role played by faces in humans.

References

Allison, T., McCarthy, G., Nobre, A., Puce, A., & Belger, A. (1994). Human exstrastriate visual cortex and the perception of faces, words, numbers, and colors. *Cerebral Cortex, 5*, 544–554.

Allison, T., Puce, A., Spencer, D. D., & McCarthy, G. (1999). Electrophysiological studies of human face perception. I. Potentials generated in occipitotemporal cortex by face and non-face stimuli. *Cerebral Cortex, 9*, 415–430.

Bentin, S., McCarthy, G., Perez, E., Puce, A., & Allison, T. (1996). Electrophysiological studies of face perception in humans. *Journal of Cognitive Neuroscience, 8*, 551–565.

Bowles, D. C., McKone, E., Dawel, A., Duchaine, B., Palermo, R., Schmalzl, L., et al. (2009). Diagnosing prosopagnosia: Effects of aging, sex, and participant-stimulus ethnic match on the Cambridge Face Memory Test and Cambridge Face Perception Test. *Cognitive Neuropsychology, 26*(5), 423–455.

Bruce, V., & Young, A., (1986). Understanding face recognition. *British Journal of Psychology, 77* (3), 305–327.

Carey, S., Diamond, R., & Woods, B. (1980). Development of face recognition—a maturational component? *Developmental Psychology, 16*(4), 257–269.

Cohen, L. B., & Cashon, C. H. (2001). Do 7-month-old infants process independent features of facial configurations? *Infant and Child Development, 10*, 83–92.

Crouzet, S., Kirchner, H., & Thorpe, S. J. (2010). Fast saccades toward faces: face detection in just 100 ms. *Journal of Vision, 10*(4), 1–17.

Deffke, I., Sander, T., Heidenreich, J., Sommer, W., Curio, G., Trahms, L., et al. (2007). MEG/EEG sources of the 170-ms response to faces are co-localized in the fusiform gyrus. *NeuroImage, 35*, 1495–1501.

Diamond, R., & Carey, S. (1986). Why faces are and are not special: An effect of expertise. *Journal of Experimental Psychology: General, 115*(2), 107–117.

Duchowny, M. S. (1989). Surgery for intractable epilepsy: Issues and outcome. *Pediatrics, 84,* 886–894.

Feingold, C. A. (1914). The influence of environment on identification of persons and things. *Journal of Criminal Law and Police, 5,* 39–51.

Gauthier, I., Tarr, M. J., Moylan, J., Skudlarski, P., Gore, J. C., & Anderson, A. W. (2000). The fusiform "face area" is part of a network that processes faces at the individual level. *Journal of Cognitive Neuroscience, 12,* 495–504.

Gobbini, M. I., & Haxby, J. V. (2007). Neural systems for recognition of familiar faces. *Neuropsycholgia, 45,* 32–41.

Grill-Spector, K., & Kanwisher, N. (2005). As soon as you know it is there, you know what it is. *Psychological Science, 16*(2), 152–160.

Gross, C. G. (2008). Single neuron studies of inferior temporal cortex. *Neuropsychologia, 46,* 841–852.

Harris, A., & Nakayama, K. (2008). Rapid adaptation of the M170 response: Importance of face parts. *Cerebral Cortex, 18,* 467–476.

Haxby, J. V., Hoffman, E. A., & Gobbini, M. I. (2000). The distributed human neural system for face perception. *Trends in Cognitive Sciences, 4*(6), 223–233.

Haxby, J. V., Horwitz, B., Ungerleider, L. G., Maisog, J. M., Pietrini, P., & Grady, C. L. (1994). The functional organization of human extrastriate cortex: A PET-rCBF study of selective attention to faces and locations. *Journal of Neuroscience, 14,* 6336–6353.

Hayden, A., Bhatt, R. S., Reed, A., Corbly, C. R., & Joseph, J. E. (2007). The development of expert face processing: Are infants sensitive to normal differences in second-order relational information? *Journal of Experimental Child Psychology, 97,* 85–98.

Itier, R. J., Alain, C., Sedore, K., & McIntosh, A. (2007). Early face processing specificity: It's in the eyes! *Journal of Cognitive Neuroscience, 19*(11), 1815–1826.

Jeffreys, D. A. (1989). A face-responsive potential recorded from the human scalp. *Experimental Brain Research, 78,* 193–202.

Jonas, J., Descoins, M., Koessler, L., Colnat-Coulbois, S., Sauvee, M., Guye, M., et al. (2012). Focal electrical intracerebral stimulation of a face-sensitive area causes transient prosopagnosia. *Neuroscience, 222*(11), 1078–1091.

Kanwisher, N., McDermott, J., & Chun, M. M. (1997). The fusiform face area: A module in human extrastriate cortex specialized for face perception. *Journal of Neuroscience, 17,* 4302–4311.

Kriegeskorte, N., Formisano, E., Sorger, B., & Goebel, R. (2007). Individual faces elicit distinct response patterns in human anterior temporal cortex. *Proceedings of the National Academy of Science USA, 104*(51), 20600–20605.

Kuhl, P. K., Tsao, F.-M., & Liu, H.-M. (2003). Foreign-language experience in infancy: Effects of short-term exposure and social interaction on phonetic learning. *Proceedings of the National Academy of Sciences (PNAS) USA, 100,* 9096–9101.

Le Grand, R., Mondloch, C., Maurer, D., & Brent, H. (2004). Impairment in holistic face processing following early visual deprivation. *Psychological Science, 15,* 762–768.

Linkenkaer-Hansen, K., Palva, J. M., Sams, M., Hietanen, J. K., Aronen, H. J., & Ilmoniemi, R. J. (1998). Face-selective processing in human extrastriate cortex around 120 ms after stimulus onset revealed by magneto- and electroencephalography. *Neuroscience Letters, 253,* 147–150.

Liu, J., Harris, A., & Kanwisher, N. (2002). Stages of processing in face perception: An MEG study. *Nature Neuroscience, 5*(9), 910–916.

Liu, J., Harris, A., & Kanwisher, N. (2009). Perception of face parts and face configurations: An fMRI study. *Journal of Cognitive Neuroscience, 22*(1), 203–211.

Liu, J., Higuchi, M., Marantz, A., & Kanwisher, N. (2000). The selectivity of the occipitotemporal M170 for faces. *Cognitive Neuroscience and Neuropsychology, 11*(2), 337–341.

McKone, E., Crookes, K., Jeffery, L., & Dilks, D. (2012). A critical review of the development of face recognition: Experience is less important than previously believed. *Cognitive Neuropsychology, 29,* 174–212.

McKone, E., Crookes, K., & Kanwisher, N. (2009). The cognitive and neural development of face recognition in humans. In M. S. Gazzaniga (Ed.), *The cognitive neurosciences IV* (pp. 467–482). Cambridge, MA: Bradford Books.

McKone, E., & Yovel, G. (2009). Why does picture-plane inversion sometimes dissociate perception of features and spacing in faces, and sometimes not? Toward a new theory of holistic processing. *Psychonomic Bulletin & Review 16*, 778–797.

McKone, E., Kanwisher, N., & Duchaine, B. (2006). Can generic expertise explain special processing for faces? *Trends in Cognitive Sciences, 11*(1), 8–15.

McKone, E., & Palermo, R. (2010). A strong role for nature in face recognition. *Proceedings of the National Academy of Sciences (PNAS), US, 107*(11), 4795–4796.

Milner, A. D., & Goodale, M. A. (2006). *The visual brain in action.* New York: Oxford University Press.

Mondloch, C., Le Grand, R., & Maurer, D. (2002). Configural face processing develops more slowly than featural processing. *Perception, 31*, 553–566.

Narumoto, J., Okada, T., Sadato, N., Fukui, K., & Yonekura, Y. (2001). Attention to emotion modulates fMRI activity in human right superior temporal sulcus. *Brain Research. Cognitive Brain Research, 12*(2), 225–231.

Ostrovsky, Y., Andalman, A., & Sinha, P. (2006). Vision following extended congenital blindness. *Psychological Science, 17*(12), 1009–1014.

Palermo, R., & Rhodes, G. (2002). The influence of divided attention on holistic face perception. *Cognition, 82*, 225–257.

Parvizi, J., Jacques, C., Foster, B. L., Withoft, N., Rangarajan, V., Weiner, K. S., et al. (2012). Electrical stimulation of human fusiform face-selective regions distorts face perception. *The Journal of Neuroscience, 32*(43), 14915–14920.

Pascalis, O., de Haan, M., & Nelson, C. A. (2002). Is face processing species-specific during the first year of life? *Science, 296*(5571), 1321–1323.

Pascalis, O., deHaan, M., Nelson, C. A., & de Schonen, S. (1998). Long-term recognition memory for faces assessed by visual paired comparison in 3- and 6-month-old infants. *Journal of Experimental Psychology. Learning, Memory, and Cognition, 24*(1), 249–260.

Pitcher, D., Duchaine, B., Walsh, V., Yovel, G., & Kanwisher, N. (2011). The role of lateral occipital face and object areas in the face inversion effect. *Neuropsychologia, 49*(12), 3448–3453.

Pitcher, D., Walsh, V., Yovel, G., & Duchaine, B. (2007). TMS evidence for the involvement of the right occipital face area in early face processing. *Current Biology, 17*, 1568–1573.

Purcell, D. G., & Stewart, A. L. (1988). The face detection effect: Configuration enhances perception. *Perception and Psychophysics, 43*(4), 355–366.

Rivolta, D., Palermo, R., Schmalzl, L., & Williams, M. A. (2012). An early category-specific neural response for the perception of both places and faces. *Cognitive Neuroscience, 3*(1), 45–51.

Robbins, R., & McKone, E. (2007). No face-like processing for objects-of-expertise in three behavioural tasks. *Cognition, 103*, 331–336.

Rossion, B., Gauthier, I., Tarr, M. J., Despland, P., Bruyer, R., Linotte, S., et al. (2000). The N170 occipito-temporal component is delayed and enhanced to inverted faces but not to inverted objects: An electrophysiological account of face-specific processes in the human brain. *NeuroReport, 11*, 69–74.

Rotshtein, P., Henson, R. N. A., Treves, A., Driver, J., & Dolan, R. (2005). Morphing Marilyn into Maggie dissociates physical and identity face representations in the brain. *Nature Neuroscience, 8*(1), 107–113.

Scherf, K. S., Behrmann, M., Humphrey, K., & Luna, B. (2007). Visual category-selectivity for faces, places and objects emerges along different developmental trajectories. *Developmental Science, 10*(4), F15–F30.

Schiltz, C., & Rossion, B. (2006). Faces are represented holistically in the human occipito-temporal cortex. *NeuroImage, 32*, 1385–1394.

Schmalzl, L. (2007). *Fractionating face processing in congenital prosopagnosia.* Sydney, Australia: Macquarie University.

Sengpiel, F. (2007). The critical period. *Current Biology, 17*, R742–R743.

Slater, A., Quinn, P. C., Hayes, R., & Brown, E. (2000). The role of facial orientation in newborn infants' preference for attractive faces. *Developmental Science, 3*(2), 181–185.

Tanaka, J. W. (2001). The entry point of face recognition: Evidence for face expertise. *Journal of Experimental Psychology: General, 130*(3), 534–543.

Tanaka, J. W., & Farah, M. J. (1993). Parts and wholes in face recognition. *Quarterly Journal of Experimental Psychology A, 46*(2), 225–245.

Turati, C., Bulf, H., & Simion, F. (2008). Newborns' face recognition over changes in viewpoint. *Cognition, 106*, 1300–1321.

Wilmer, J. B., Germine, L., Chabris, C. F., Chatterjee, G., Williams, M., Loken, E., et al. (2010). Human face recognition ability is specific and highly heritable. *Proceedings of the National Academy of Science.*. doi:10.1073/pnas.1000567107.

Yin, R. K. (1969). Looking at upside-down faces. *Journal of Experimental Psychology, 81*, 141–145.

Young, A. W., Hellawell, D., & Hay, D. C. (1987). Configurational information in face perception. *Perception, 16*, 747–759.

Yovel, G., & Kanwisher, N. (2005). The neural basis of the behavioral face-inversion effect. *Current Biology, 15*, 2256–2262.

Chapter 3
Prosopagnosia: The Inability to Recognize Faces

Most people recognize familiar faces rapidly, accurately and effortlessly. However, this is not true for individuals with prosopagnosia, who show a deficit in recognizing familiar people by their faces.

The word prosopagnosia, from the Greek *prosopon* (face) and *a-gnois* (without knowledge), was coined for the first time by the German neurologist Bodamer (1947) who described three cases, including a 24 year old man who lost his ability to recognize faces after suffering a bullet wound to his head. In contrast, he was able to identify familiar people through other sensory modalities such as hearing and touch, or even through extra-facial visual cues such as gait and physical mannerisms. This case reflects an acquired condition, that is, a condition that was not present from birth but that followed some traumatic events, thus it is known as *acquired prosopagnosia*. In the last 20 years, however, researchers also focused their attention on another form of prosopagnosia that appears without any evident cause, and which is known as *congenital prosopagnosia*.

In a nutshell, people suffering from acquired prosopagnosia have lost the ability to properly recognize faces, which was absolutely normal before a neurological event, whereas people with congenital prosopagnosia have never developed typical face recognition abilities despite the absence of any evident neurological problem. In this chapter we will learn about both forms of prosopagnosia and about their similarities and differences. Both conditions represent unique opportunities for cognitive science to understand the normal cognitive and neural aspects of face processing.

3.1 Acquired Prosopagnosia

3.1.1 History of the Condition

Michael is a 62 years old man who suffered a stroke while he was fishing at a lake in a small village not too far from Sydney. Peter, his son, who luckily was with him in that moment, promptly transferred his father to the closest hospital. A few hours later the doctor gave the bad news to the family: Michael had had

D. Rivolta, *Prosopagnosia*, Cognitive Systems Monographs 20,
DOI: 10.1007/978-3-642-40784-0_3, © Springer-Verlag Berlin Heidelberg 2014

stroke in the right hemisphere of his brain. Two months later Michael underwent a neuropsychological examination to evaluate his cognitive functions. His language, memory, attention and the capacity to recognize objects were largely intact, however, Michael was no longer able to memorize new faces nor to recognize familiar faces. He could only recognize his wife and his son by using extra-facial tools, such as the hair-style or a particular outfit. It has been also reported that on some occasions Michael could not recognize himself in the mirror or even in pictures taken a few months before the injury. By using his own words: "My life changed completely; I feel so uncomfortable and ashamed of my condition that I rather stay at home avoiding meeting friends that, certainly, I cannot recognize".

Michael suffers from acquired prosopagnosia. Public awareness of this particular condition rose in the 80s by the beautiful descriptions provided by Dr Oliver Sacks, who described in his book the case of a "man who mistook his wife for a hat", due to a severe impairment in visual recognition.

The first observation that face recognition can be impaired in brain injured individuals can be traced back to the ancient Greeks (Thucidydes II, 49–50), with reports of soldiers injured in the Peloponnesian War who exhibited "strange behaviours" including severe memory problems and an inability to recognize friends (Schmalzl 2007).

Jumping back to the nineteenth century, similar difficulties were reported in patients who suffered neurological diseases. The first scientific observation of a patient's inability to recognize familiar faces comes from Wigan who, in 1844, described a man with a complete impairment in remembering faces. Other descriptions came from Quaglino (1867), Jackson (1872), Charcot and Bernard (1883), and Wilbrand (1892) who described patients with serious impairments in perceiving, remembering and recognize faces; even very familiar faces and sometimes their own face in the mirror (see Quaglino et al., 2003, for a discussion on the topic). At the time it was not possible to conduct non-invasive examinations of the brain anatomy after the injury (e.g., structural MRI); however the clinical features of the patients who typically reported left hemiplegia[1] led some investigators, such as Quaglino, to argue for right-sided brain lesions. A few decades later, Hoff and Poetzl (1937) explicitly suggested that face recognition might be a separate function that can be impaired following brain injury (for a more detailed description of these cases please refer to Della Sala and Young 2003; Schmalzl 2007).

However, it was not until 1947 that Bodamer described the inability to recognize familiar faces to be a selective form of visual agnosia,[2] referring to it as "prosopagnosia". Since Bodamer's seminal paper, more than a hundred case reports of prosopagnosia have been published (De Haan 2001).

---

[1] The word hemiplegia refers to the inability to move half side (the contro-lesional one) of the body.

[2] In neuropsychology visual agnosia is the name of a general condition characterized by the inability to recognize visual stimuli despite normal low-level vision (e.g., normal colour perception and normal visual field). Patients can show inability to recognize objects (*object agnosia*), colours (*colour agnosia*) and/or faces (*prosopagnosia*) (Denes and Pizzamiglio 1996).

3.1.2 Causes (Aetiologies) of Acquired Prosopagnosia

In general, cognitive deficits such as acquired prosopagnosia can have very different aetiologies. One of the most common is the *Stroke* (or cerebrovascular accident), that refers to the disruption of blood flow in the brain which is a consequence of blood vessel occlusion (ischemic stroke) or perforation (hemorrhagic stroke). When blood flood in the brain is diminished after a stroke, brain cells (i.e., neurons) do not receive the oxygen and glucose they need to function and die (Kempler 2005). When not lethal, this event can give rise to different cognitive deficits according to the size and the location of the lesion. If the stroke involves posterior blood vessels it can cause prosopagnosia (Lang et al. 2006).

In traumatic events such as car accidents and sports injuries, the head can receive a strong trauma (*head injury*) that can cause neuronal loss leading to prosopagnosia (Farah et al. 1995). Acquired prosopagnosia can even be a consequence of surgical lobectomies for the treatment of epilepsy (Braun et al. 1994), degenerative disorders (Williams et al. 2006) and carbon monoxide poisoning (Sparr et al. 1991).

3.1.3 Clinical and Neural Features of Acquired Prosopagnosia

Acquired prosopagnosia, albeit caused by very different neurological conditions, share some common features in all patients that suffer from it, with the main one being the failure to recognize other people by their face. However, how bad are they at this? Can they recognize some people? What are the features of their face recognition deficit? Is this the only specific function they have lost? What are the neural features of the condition?

In general, patients with acquired prosopagnosia are unable to experience any sense of familiarity when viewing faces of known people and they fail to identify these faces in light of (relatively) intact memory, intelligence, low-level vision, object and place (e.g., monuments) recognition. They have problems in gender identification of faces, in matching unfamiliar faces and in remembering already seen faces (Barton 2008b; Gainotti and Marra 2011; Sergent and Signoret 1992). It appears clear that the clinical features of prosopagnosia are related to the location and dimension of the neural loss. The first descriptions of the neuroanatomical underpinning of prosopagnosia underlined bilateral posterior (occipito-temporal) brain lesions (Damasio et al. 1990; Meadows 1974; Steeves et al. 2006). Over the years however there have been reports of patients with unilateral right lesions (Benton 1990; Sugimoto et al. 2011; Wada and Yamamoto 2001) and very few with left lesions (Mattson et al. 2000).[3]

---

[3] It is not the aim of the chapter to provide an extensive review of the anatomical and clinical features of cases of acquired prosopagnosia. The interested reader should refer for example to Barton (2008) and Gainotti (2011).

The face recognition system is mediated by different anatomical sources spanning from posterior to most anterior brain regions. According to the Haxby model (see Chap. 2), the first area involved in face processing is the OFA that, in turn, will send information to FFA and other areas. From this view it follows that a selective lesion involving the OFA will cause abnormal functioning of all other face areas. One way to test this hypothesis is by investigating in detail single case studies of prosopagnosics using human neuroimaging. Over the last 10 years Rossion and collaborators provided extensive reports of PS, a woman born in 1950 that sustained a severe close-head injury in 1992 after being hit by a bus. This accident mainly caused a brain lesion in the lateral part of the occipito and temporal lobes bilaterally. Even many years after the accident and after neuropsychological rehabilitation PS still presents a massive prosopagnosia that prevents her from recognizing famous and familiar people by their faces. This impairment contrasts with her excellent person recognition from people's voice, hairstyle, gait, size and posture (Rossion et al. 2003). For example, she could feel a sense of familiarity only for 14 faces out of 60 that she knows really well by their names, and she could recognize only four of them. Despite this difficulty with face recognition, PS did not report any complaints about object recognition that, as demonstrated by neuropsychological tests, is well within the normal range (see a detailed summary of PS' investigations in Rossion 2009). The interesting aspect about PS's case is that she has a brain lesion that affects the normal activity of the right OFA, sparing the more anterior FFA. In other words, by using fMRI it is possible to see the FFA, but not the OFA. This suggests that visual information can reach the FFA, where neural activity categorizes the perceived stimulus as a face, without receiving any information from the OFA, areas where featural processing is taking place. This and further experimental evidence (Ewbank et al. 2012) posits against a hierarchical proposal of visual areas, and suggests that the OFA plays a crucial role for normal face processing (see Fig. 3.1).

PS's case represents an example of how a single-case detailed analysis conducted at a behavioural and neural level can provide precious information about normal and abnormal aspects of face recognition. It is not the intention of the current work to discuss the likelihood of different models of face recognition; this topic, by itself, may require an entire book.

Without spending further time in dealing with theoretical questions regarding the recruitment order of face areas, studies in acquired prosopagnosia have demonstrated:

- Left side (posterior) lesions are often characterized by a general visual agnosia that can involve faces (in this case prosopagnosia is just a symptom of a more general agnosic problem.
- Right sided (posterior) lesions cause prosopagnosia as a distinct syndrome characterized by the impairment in familiarity feelings (i.e., telling whether the seen face is familiar or not) and holistic processing (i.e., processing the spacing between features).

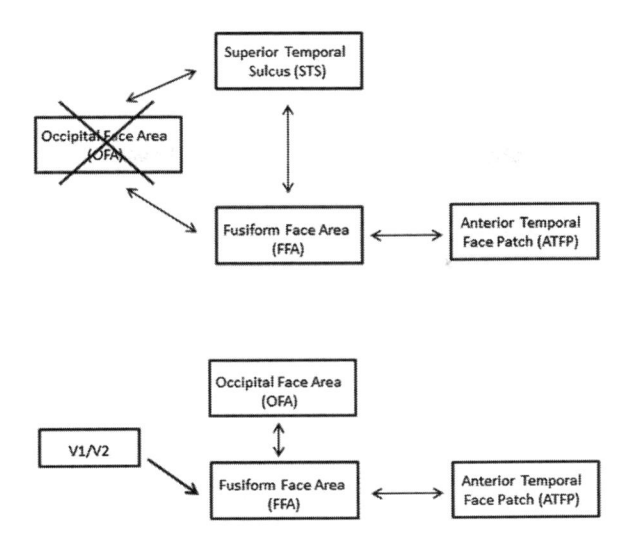

Fig. 3.1 Rossion's proposal for the organization of face areas. Patient PS, who shows no activity of OFA (impairment indicated with an "X", see *top* of the figure) still shows normal FFA activity. Since this is in disagreement with the hierarchical model proposed by Haxby, Rossion et al. suggest that visual information coming from early visual areas (V1/V2) reaches first the FFA, and then, OFA, before reaching identity specific regions in the anterior temporal lobe

- Right anterior temporal lesions particularly affect face imagery (i.e., describe the facial features of Lady Diana's face, as recollected from memory) and the recognition of familiar faces (Barton 2008a; Gainotti and Marra 2011).

3.1.4 Is Acquired Prosopagnosia Affecting Only the Processing of Faces?

Face recognition represents, by definition, the main problem of people affected by acquired prosopagnosia. However, it is important to underline that some patients, especially the ones who even show a left hemisphere involvement, can report deficits that are not restricted to face processing. For instance, some patients can report problems in recognizing objects such as cars, fruits and vegetables (object agnosia) and others can be impaired in decoding facial expressions. This heterogeneity is mainly driven by variability in the brain lesions. Since brain regions involved in face, objects, places and body recognition are close to each other on the cortical sheet (Kanwisher 2010), it is likely that a brain lesion will involve one, some or all of them. The specificity of face recognition impairments depends on whether the lesion is enough small and localized to just affect face-specific cortical areas such as OFA or FFA sparing other regions such as the lateral occipital complex, or the extrastriate body area. For example patient PS described above had normal object recognition, whereas patient DF (Steeves et al. 2006) showed impairments in this ability. Research with fMRI indicated that this may be the case because PS showed a normal lateral occipital-complex (LOC, Malach et al. 1995) activity, the area that mediates object processing, whereas DF did not (Steeves et al. 2009).

It is, however, important to note that even though the majority of acquired prosopagnosia cases reported in the literature evidenced problems with non-face stimuli, they were usually much less severe than face processing deficits. There are many cases where, even after a complete cognitive assessment, face-related tasks impairment was the only impairment detectable. For example the case of one patient was described who, after becoming prosopagnosic and unable to recognize human faces, could still recognize all the sheep of his breed. Another patient, who developed prosopagnosia after a car accident, was unable to recognize his closest relatives and even his face at the mirror, but was able to recognize many different objects, different exemplars of objects of the same category (glasses) and was more accurate in perceiving inverted faces than upright faces (Farah 1996).

A recent case of a 30 years old girl who has suffered from an embolization of a right posterior inferior arteriovenous malformation that affected occipito-temporal brain regions, reported a selective problem in recognizing familiar faces out of an expected context that tried to circumvent by focusing on clothing, hair-style and voice (Riddoch et al. 2008). A detailed neuropsychological assessment reported serious problems in face recognition since she recognized none out of thirty faces that were familiar before the injury, she had problems in matching faces that differ in their orientation and she showed a lack of holistic process-ing. Despite this, her ability to identify emotions, age and gender from faces, her intelligence, object recognition skills and general memory were all well within the normal range.

Acquired prosopagnosia does not only involve adults. Adam, a 16 year old boy, developed streptococcal meningitis at one day of age,[4] suffering infarction of the posterior cerebral arteries and sustaining bilateral occipito-temporal lesions (Farah et al. 2000). At the time of investigation, Adam showed severe face recognition impairments, and while his object recognition was not flawless, it was considera-bly better than his face recognition, and fully adequate for activities of everyday living. The authors of the study pointed out that the fact that other brain systems did not take over the functions supporting normal face recognition suggests that specialized processing systems for face recognition exist even prior to visual expe-rience. This in turn suggests that the functional delineation and anatomical

---

[4] The fact that the brain lesion was acquired short after birth, makes Adam a case of develop-mental prosopagnosia (i.e., prosopagnosia acquired very early in development) and not of con-genital prosopagnosia (see below).

localization of face recognition is at least to some extent specified in the genome (see Chap. 2).

Cases of acquired prosopagnosia, together with the complementary case of people showing only object recognition deficits, *object agnosia*, but normal face processing (Moscovitch et al. 1997), strengthens what we learned in the previous chapter; that faces represent a class of special stimuli for our visual system that are mediated by specialized neural architecture.

3.2 Congenital Prosopagnosia

3.2.1 What is it and Why do We Study it?

MJ, a seven-year-old boy, was an enigma for his mother. At home or when playing with unfamiliar children in public places, he was friendly and engaging. At school though he was a loner who tended to watch the other children play. His teachers regularly became exasperated with him when he failed to follow their instructions. MJ refused his teacher's request to take papers to a particular student, claiming that he didn't know who the student was. When told to stand next to a student names Casey, MJ stood next to the student he thought was Casey but his teacher became angry and she sent him to the end of the line. MJ's mother's claims that he was normal at home were met with disbelief by his teachers, and MJ's principal told his mother that his social problems were probably caused by her anxiety. There were other oddities as well. MJ wasn't able to recognize the small number of neighbors they met regularly. [...] In preschool, the only student MJ could identify was a Chinese girl—the only non-caucasian student in the class. [...] MJ's one friend in kindergarten was a boy named Jacob. Upon running into Jacob and his parents at a soccer game, MJ failed to recognize him. The adults laughed off MJ's inability to recognize Jacob, but MJ's mother could see the same confused look on his face she'd seen many times before—a lack of recognition coupled with the look of someone who thinks they're being duped. Then, a couple of week later, MJ's older brother saw one of his former classmates at church. His name was also Jacob and he was roughly the same size and coloring as MJ's friend Jacob. Upon seeing his brother talking to this Jacob, MJ clearly though this boy was his friend Jacob. His mother was astounded, and the realization hit her like a ton of bricks. MJ could not recognize people! (from Duchaine 2011).

From this description it appears that MJ suffers from congenital prosopagnosia. Since the first case study description of congenital prosopagnosia, a 12 years old girl who reported having severe difficulties recognizing faces that she was not extremely familiar with (McConachie 1976), many laboratories extensively investigated face processing in this condition.

One of the main differences between acquired and congenital prosopagnosia is that the former, given its traumatic origin, gets immediately recognized, whereas the congenital form can get undetected for a long time as the individual has no means of comparison with normal face processing skills. Also, because congenital prosopagnosics have had a lifetime to develop compensatory strategies, they are adept at using salient features such as hairline or eye brows for recognition.

Given the absence of any evident sign of brain lesion, its developmental nature and its face-specificity (i.e., face recognition very often represents the only identifiable problem), congenital prosopagnosia represents a unique opportunity for the investigation of the cognitive and neural aspects of normal face processing. Furthermore, congenital prosopagnosia allows us to explore the necessary and sufficient brain activations that are required to ensure intact face recognition. In addition, given that CP individuals never develop normal face recognition despite seemingly normal exposure to faces throughout their life, congenital prosopagnosia raises numerous theoretical questions regarding the psychological and neural conditions required for the initial acquisition of normal face processing and possible limitations on the time period during which this acquisition can occur. Intriguingly, congenital prosopagnosia also serves as a model to explore whether, and to what extent, face processing is amenable to target intervention programs and, if so, what underlying neural changes might serve as the correlate accompanying the behavioural improvement (Schmalzl 2007).

Before providing further details on congenital prosopagnosia, it is important to clarify some potentially confusing aspects regarding the names with whom this condition has been referred to. Some researchers refer to congenital prosopagnosia using the name *developmental prosopagnosia* (Duchaine et al. 2006; Kress and Daum 2003). Both congenital and developmental prosopagnosia have often been used as synonyms, to indicate people with difficulties in face recognition *from birth* without any obvious evidence of any neurologic and/or psychiatric disorder. I have to note however that some researchers adopted the term developmental prosopagnosia to indicate people who developed prosopagnosia as a consequence of sustained brain damage before, during or immediately after birth, i.e. prior to any visual experience with faces (Barton et al. 2003). In this book, in line with other researchers (Behrmann and Avidan 2005), I refer to these cases as developmental prosopagnosia, whereas cases of prosopagnosia from birth without any history of brain damage as congenital prosopagnosia.

3.2.2 Behavioural Features of Congenital Prosopagnosia: Towards the Diagnosis

How is it possible to identify congenital prosopagnosia? In order to study, and potentially treat it, it is first necessary to identify some features that can constitute a diagnosis for this condition. Despite the existence of a general consensus on the fact that the core feature of congenital prosopagnosia is an utter failure in the identification of familiar faces, there are still not standard diagnostic criteria that virtually all people in the field can follow. Luckily however, there are some very well done behavioural tasks to use and some general criteria that must be considered in order to formulate a diagnosis of congenital prosopagnosia.

People can be considered as suffering from a "pure" form of congenital prosopagnosics if they show:

- Impairment on behavioural tasks that measure face processing skills such as face memory, face recognition and face perception;

- Normal, or relatively normal, performance on tasks that measure the memory, recognition and perception of non-face visual stimuli such as objects;
- Normal general intelligence (IQ);
- Normal overall cognitive functioning such as normal memory and attention;
- Normal social skills (i.e., no autistic traits);
- Normal low-level vision (e.g., contrast sensitivity, acuity, colour perception);
- Absence of evident signs of neurological (e.g., epilepsy) and/or psychiatric (e.g., psychosis) conditions.

We start now with a description of these points in more detail by indicating real tasks that psychologists use for the identification of people suffering from congenital prosopagnosia. Note that these tools are not specific for congenital prosopagnosia, but can be well adopted if it is suspected that, for instance, face processing impairments are due to ictus, epilepsy, head trauma or Alzheimer disease.

3.2.3 Performance on Behavioral Tasks

As indicated by the Bruce and Young model of face recognition described in the previous chapter, normal face perception is not a monolithic phenomenon, but represents a variegated process in which visual information gets elaborated over different stages. For instance, face processing requires the detection of a face among other types of visual stimuli, the discrimination between different faces even if they are unfamiliar, and finally, the recognition, in which the exact identity of an individual faces is extracted. Since each of these stages can be selectively targeted by a congenital or acquired condition, thus leading to many potentially different behavioural features of the face recognition impairment, it appears clear that the definition of what we mean for face processing deficit is a tricky one. Although congenital prosopagnosia has recently attracted much scientific attention, many aspects of its cognitive profile and underlying neural functioning are still unclear.

Albeit most scientists would consider the inability to recognize familiar faces as the hallmark of the disorder, the extent of the impairment in other face- and non face-processing is still far from clear. Investigations in congenital prosopagnosia conducted over the last 10 years show an abundance of *heterogeneity*; this means that different subjects can have different "phenotypes",[5] that is, they can show different cognitive, but also neural and neurophysiological, profiles.

Different laboratories over the world choose what they believe to be the most appropriate battery to detect prosopagnosia. Below I report the tasks we adopted at the Macquarie Centre for Cognitive Science (MACCS, Sydney, Australia). They include tasks aimed at assessing face related skills and tasks aimed at tapping into non-face processing and general intelligence. Below I start reporting three

---

[5] The term "phenotype", borrowed from Genetics, indicates here the strengths and weaknesses that characterize face recognition skills in different people with congenital prosopagnosia.

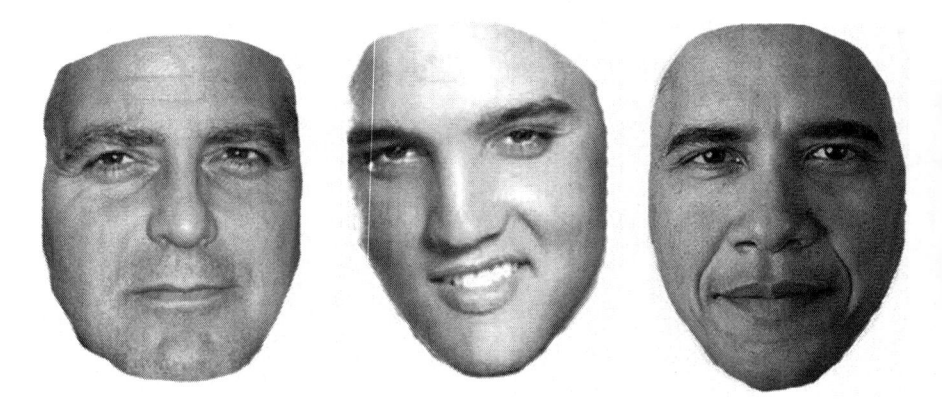

Fig. 3.2 Can you recognize these three faces? If not, but you are very familiar with them, (Starting from the left, these faces are George Clooney, Elvis Presley and Barack Obama) you may have some face recognition difficulties. The presentation of faces without hair represents a good way to test for the recognition of famous faces since external clues such as hair style cannot cue person identification

important tasks frequently reported by researchers in the field: a task that assesses the recognition of famous faces, a task that assesses face memory for unfamiliar faces and, finally, a task that investigates face perception skills.

3.2.3.1 Recognition of Familiar Faces

The problem people with congenital prosopagnosia experience in everyday life concerns the inability to recognize familiar/known faces. Thus, one of the mostly adopted ways to investigate face recognition skills is by presenting personally familiar or/and famous faces. Since the collection of pictures of personally familiar faces, despite representing a very valid approach, requires a lot of effort from both the researchers and family members' sides, the great part of investigations in congenital prosopagnosia adopted famous faces,[6] such as politicians, actors, sport people, models and so forth. Famous faces can be presented with hair or without hair. This second option has to be preferred because it is more sensitive, since it does not allow participants to adopt strategies that focus on non-face external cues, such as hairstyle, to mediate their face recognition. The use of external cues can in fact boost the performance of people that actually have problems in recognizing faces (see Fig. 3.2).

On a task, called the *MACCS-08 Famous Face Test* (or MFFT-08), participants are presented with 20 famous and 20 unfamiliar faces (Palermo et al. 2011a). On each trial (i) a face is presented and participants judge whether it is familiar or not, (ii), for the famous faces, they are asked to identify the face by providing its name or other specific

---

[6] The main problem in adopting personally familiar faces is that we need a strong collaboration from family member who should collect many pictures of family members and friends of the prosopagnosic. This, unfortunately, is not always possible.

autobiographical information, then (iii) the famous person's name and relevant autobiographical information is provided and participants report whether the famous person was actually known to them (the unknown where excluded from the analysis).

The score on the MFFT-08 is the percentage (%) of correctly recognized faces of known famous people. Like on all other neuropsychological tasks, before telling whether a participant performs within the normal range, we need to collect data from healthy volunteers, that is, people that do not report any problems in face recognition and do not show any psychiatric and/or neurological symptoms. Then, using statistical tests we compare the specific participant to the normal population. Sometimes there are sex and/or age differences in the performance of the normal population on a specific task, so that, for instance, females are better than males or vice versa. In these situations it is very important to consider the gender and the age of our potential prosopagnosic before referring to him as a prosopagnosic or not (Bowles et al. 2009; McKone et al. 2011).

3.2.3.2 Face Perception Skills

The ability to perceptually discriminate between similar faces can be assessed by the *Cambridge Face Perception Test* (CFPT, Duchaine et al. 2007). On each trial, participants are required to order a series of sixed morphed front-view faces in term of their similarity to a target face photographed from a three-quarter view (see Fig. 3.3).

3.2.3.3 Memory for Unfamiliar Faces

A task that is receiving much attention because of its good psychometric characteristics is the *Cambridge Face Memory Task* (CFMT, Duchaine and Nakayama 2006a) (see Fig. 3.4).

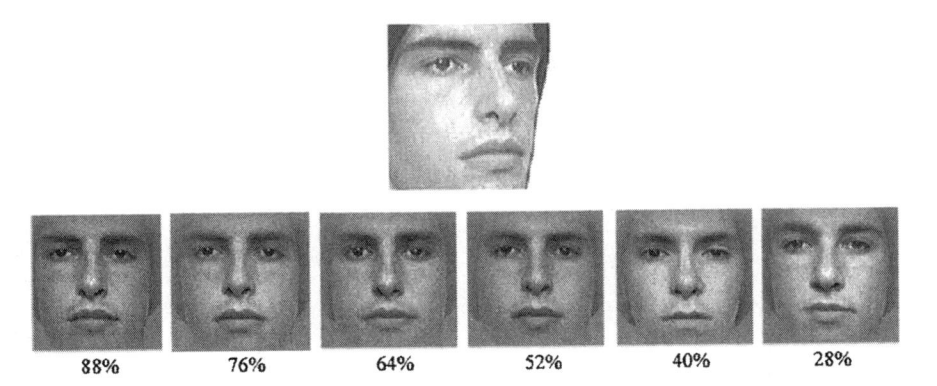

88%   76%   64%   52%   40%   28%

Fig. 3.3 The Cambridge Face Perception Task (CFPT). Faces have to be sorted starting from the most similar to the target, to the last similar. The percentage below each face indicates the similarity between that face to the target

Fig. 3.4 The Cambridge
Face Memory Task (CFMT).
Here is represented one
identity out of six to learn
(*top raw*). Subjects are
asked to identify the learnt
faces amongst distractors
in three different sequential
conditions: same orientation
(*second raw*), different
orientation and light (*third
raw*) and visual noise (*bottom
raw*)

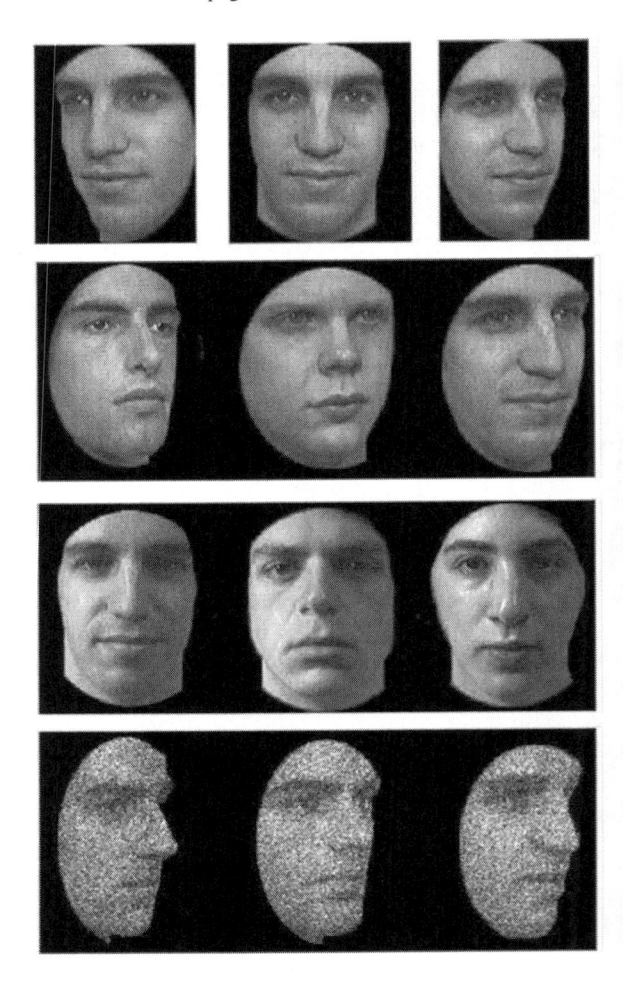

The CFMT assesses face learning and memory. Participants learn the faces of six people they have never seen before and then select the learnt faces from two similar distractors on trials in which the test faces are either identical to the learnt images, varied in lighting and viewpoint, or degraded with visual noise.

3.2.4 Other Behavioural Characteristics of Congenital Prosopagnosics

These three tasks have been extensively used as diagnostic tools for congenital prosopagnosia research. For example people can be included in the study as "prosopagnosics" if they show abnormal performance (i.e., two standard deviations below the controls mean) on at least one of these three tasks. In the typical scenario, however, the performance of people with face recognition difficulties is below the normal range on two or even all of them (Rivolta 2010).

It is important to note that many other face tasks have been used in congenital prosopagnosia research; these are not, in general, adopted for diagnostic purposes, but are mostly adopted for theoretical reasons (e.g., investigation of holistic processing). For instance, people with congenital prosopagnosia can typically acknowledge that the shown stimulus is a face and not another object, and they can often identify the gender, the approximate age and attractiveness of faces. Even emotion recognition is typically within the normal range in congenital prosopagnosia, confirming the subjective reports of everyday-life specific problems in facial identification, but non expression, (Duchaine et al. 2003; Humphreys et al. 2007; Palermo et al. 2011b). We know by now that typical face processing is mediated by holistic processing. The importance of these mechanisms for typical face processing is strengthened by cases of acquired prosopagnosia who show the absence of those mechanisms (Busigny et al. 2010). Similarly to what is described in acquired prosopagnosia literature, different cases of people suffering from congenital prosopagnosia, with few exceptions (Susilo et al. 2011), show abnormal holistic processing as indicated by the lack of the classic face inversion effect, the absence of the composite face effect and the impairments in detecting the distance between features (e.g., the Jane Task) (Behrmann et al. 2005; Rivolta et al. 2010).

3.2.5 Non-face Processing Assessment

To formulate a "diagnosis" of congenital prosopagnosia it is important to consider whether non-face processing is within the normal range. Even though the number and the variety of these tasks can differ from study to study, a basic assessment[7] should contain tasks that investigate general intelligence (IQ), low level vision such as colour perception, contrast sensitivity, near/far vision, and object recognition. The problems in face recognition must not be a consequence of colour blindness, basic visual difficulties and/or mental deterioration. In addition, since it has been indicated a relatively high incidence of face recognition problems in people with autism spectrum disorder (Wilson et al. 2010a, b), it is important to ascertain whether the social skills of potential congenital prosopagnosics are within the normal range. This is the reason why researchers often adopt questionnaires that, despite not being diagnostic tools for autism, are very good indices of the presence of some autistic traits (Baron-Cohen et al. 2001).

To understand whether the problem in memory for faces is specific for faces and does not represent a general memory problem, it is advisable to adopt tasks that measure non-face memory skills, for example by using cars instead of faces

---

[7] The tasks I present are the ones adopted (at least from the 2007 until 2011) at the Macquarie Centre for Cognitive Science in order to understand whether a person suffers from congenital prosopagnosia.

(Dennett et al. 2011). An important marker for the presence of congenital prosopagnosia is that participants should perform within the normal range on this task but *not* on the face version of the task. The last aspect to consider is characterized by the exclusion of neurologic and or psychiatric problems. For this reason participants are typically interviewed by experts that ascertain the potential presence of potential clinical psychiatric symptoms and, when possible, exclude the presence of obvious brain abnormalities as assessed with MRI.

Despite face recognition representing the key symptom of congenital prosopagnosia and the fact that most congenital prosopagnosics show specific and selective problems in face processing, there are some people suffering from this condition that report other problems. For instance, some have problems in spatial navigation and orientation; others have problems in judging facial expressions, attractiveness, gender and trustfulness of faces. Some, despite normal basic object recognition, have trouble with individual item object recognition (within-class object recognition), that is, they can clearly identify a table and a chair, but some are slower and less accurate than normal people in identifying different types of tables and different types of chairs (Behrmann et al. 2005; Duchaine 2011). This variability suggests that different cognitive, neural and, potentially, genetic mechanisms are involved in the condition. It is important to note that, when present, non-face processing difficulties represent a much smaller problem compared to the identification of faces, and typically, they do not reach an alarming threshold which can compromise the daily life of the person. In other words, even if diagnostic tests reveal a problem in object identification, the patient complains about a serious incapacity, for instance, to follow characters in the movies, and not to recognize and use objects around them.

It is important to note that, despite the fact that research tends to consider congenital prosopagnosia as a unique condition, behavioural studies underlie great performance variability. Thus, according to the structure of the Bruce and Young model, cases of congenital prosopagnosia can be classified as being either "apperceptive" (e.g. De Haan and Campbell 1991) or "associative" (e.g. Jones and Tranel 2001), depending on whether the individual showed a deficit in structural encoding or not respectively. Other researchers use the term "prosopamnesia" to indicate a subtype of prosopagnosia characterized by intact face perception (e.g., they can tell when two images are of the same person), but with specific problems in memorizing faces (Williams et al., 2007). Despite the theoretical importance these distinctions have, for simplicity reasons, in the rest of the book we will refer to CP as a single condition, without referring to specific subtypes.

3.2.6 Prevalence of Congenital Prosopagnosia

Until recently, congenital prosopagnosia was believed to be an extremely rare condition, with only a few published case studies before the year 2000 (Kress and Daum 2003). However, a very rapidly increasing number of congenital prosopagnosia cases have been described during the past few years (Duchaine and Nakayama 2006b; Le Grand et al. 2006), and a recent study estimated the prevalence of the condition to be around 2.5 % of the general population (Kennerknecht et al. 2006). Kennerknecht and colleagues performed a large-scale questionnaire based screening of 750 individuals from secondary schools and Muenster (Germany) University's medical school. Out of the 689 students who completed the screening questionnaire, 43 were rated as being "highly suspicious" of having congenital prosopagnosia and underwent an in depth semi structured interview. Based on their answers on the interview, 17 of them (6 males and 11 females) were confirmed to be affected, which constituted 2.5 % of the initially screened sample. Even though the main limitation of this study is that none of the individuals underwent a formal assessment of their face processing skills, a recent study that investigated face processing in 241 individuals using standard tasks (CFMT, CFPT) indicated that congenital prosopagnosia could be diagnosed in 2–2.9 % of them (Bowles et al. 2009). These two studies agree that the prevalence of congenital prosopagnosia is between 2–3 % of the general population. It may be surprising to note that even an estimated prevalence of 2 % means that congenital prosopagnosia affects, for instance, as many as 2 million people in Germany, 400,000 people in Australia, 1.2 million people in the UK and 6 million people in the US!

Congenital prosopagnosia may have often been undetected for a variety of reasons in the past. In contrast to individuals who acquire their prosopagnosia in adulthood, people with congenital prosopagnosia may be unaware of their face recognition impairments, as they are born with the condition and thus have no means of comparison with normal face processing abilities. They may simply learn to adjust to their difficulties and develop compensatory strategies for recognizing people in everyday life. Also, in contrast to individuals with prosopagnosia following acquired brain injury, people with congenital prosopagnosia are less likely to be in contact with medical practitioners and neuropsychologists who assess face recognition abilities and diagnose impairments thereof. In addition, since face processing difficulties often impact on the development of social skills (Schultz 2005), children with congenital prosopagnosia may be misdiagnosed as having behavioural problems or even autistic tendencies (Dalrymple et al. 2013). Until recently there has been very little public awareness of congenital prosopagnosia as such. Now they share their stories on the Internet through various email lists and forums, thus leading to increased awareness.[8] From the perspective of a person with congenital prosopagnosia: *"Until recently I believed it was some aberrant aspect of my personality, rather than something that has a name and a history"* (case reported in Schmalzl 2007).

At this point, a spontaneous question arises: How many kids, all over the world, might have received an erroneous diagnosis of autism or of another neurodevelopmental disorder? And again, how many of these kids isolated themselves because

---

[8] Now people can learn about prosopagnosia and share their stories, for instance, in: "http://www.faceblind.org" and http://www.maccs.mq.edu.au/research/projects/prosopagnosia/.

they were afraid to fail to recognize their friends without anyone knowing about congenital prosopagnosia? It seems clear that public awareness on the condition still has a long way to go.

3.2.7 The Genetic Basis of CP

Even though potential candidate genes remain to be established, there is increasing evidence demonstrating a genetic contribution to congenital prosopagnosia. The first suggestion of a possible familial transmission from mother to daughter was put forward by McConachie (1976), who described AB, a 12 year old girl with congenital prosopagnosia, whose mother claimed also to have prosopagnosia. However, no formal assessment was performed to confirm whether AB's mother was in fact affected. In the subsequent decades, an increasing number of individuals with congenital prosopagnosia described in the literature mentioned first-degree relatives who also had face recognition difficulties. However, only in the past couple of years have researchers begun to systematically investigate the pedigree segregation pattern of face recognition impairments in extended families. (Behrmann et al. 2005; De Haan 1999; Duchaine et al. 2007; Duchaine and Nakayama 2005; Grueter et al. 2007; Lee et al. 2010; Schmalzl et al. 2008a).

In the first of these studies (Grueter et al. 2007), 38 individuals with suspected face recognition difficulties were recruited from seven families using a screening questionnaire, and eight of them agreed to undergo a more formal (though not very detailed) assessment of their face processing skills. As a group, these eight individuals belonging to four separate families scored significantly below an age and education matched control group on a standardized test of face memory (Warrington Recognition Memory Test for Faces; Warrington 1984) as well as on a famous face recognition test. In terms of the pedigree segregation pattern, face recognition abnormalities were found to be regularly transmitted from generation to generation in all seven of the studied families, with both men and women being equally affected. The authors argued that such a segregation pattern would be best explained by a simple autosomal dominant mode of inheritance, which means that if at least one of the parents is prosopagnosic, the kids, whether they are male or female, has a 50 % chance of being prosopagnosic.[9]

In a second study, Duchaine et al. (2007) performed extensive behavioural testing with ten members (seven siblings, their parents and a paternal uncle) of a family, all exhibiting significant face recognition impairments. None of the tested family members (aged between 23–66 years) had a history of brain injury or any other neurological condition, and all had normal low level vision and general

---

[9] Humans have 23 pairs of chromosomes. The one which determines the sex, the so-called 'sexual chromosomes', are XX in the women and XY in men. The other 22 pairs are called 'autosome' or 'non-sexual chromosomes'.

intellectual functioning. They also performed within the range of controls in terms of basic configural processing (i.e. the encoding of both the global shape and local elements of compound visual stimuli), as well as on a challenging facial expression recognition task. In contrast, the family members showed significant difficulties in recognizing famous faces, learning and remembering new faces (CFMT), and perceiving the similarity between unfamiliar faces (CFPT). In addition, they showed various degrees of impairment on tasks requiring within class discrimination of objects (cars and guns). Duchaine et al. results strongly suggest that the family members' impairments result from a genetic condition affecting the development of neurocognitive mechanisms necessary for high level face as well as object recognition, and that the pedigree segregation pattern within the family may be accounted for by Grueter et al. autosomal dominant inheritance hypothesis.[10]

In sum, both of the above mentioned studies favorably support the evidence that there is a genetic contribution to CP. In addition, Duchaine et al. (2007) explicitly pointed out how within their studied family there was variability in the severity as well as the selectivity of face recognition impairments (as prosopagnosic family members also showed various degrees of impairment with within category object recognition), which in turn suggests that there is individual variability in terms of the genetic involvement. What neither of the studies has specifically addressed is whether genetically-based CP is a homogeneous phenotype with respect to the individual processing steps of face perception that can be selectively impaired. That is, deriving an intact percept of a face relies upon the integrity of a series of processing steps such as the detection of first order relations that define faces (i.e. two eyes above a nose and a mouth), and holistic processing (i.e. integrating facial features into a whole gestalt). Data from individual cases of CP (Le Grand et al. 2006) indicates that CP is a heterogeneous condition, and that the impairment in face recognition cannot be predicted by poor performance on any single one of these measure of face processing.

However, how heterogeneous is CP within the same family? Schmalzl et al. (2008a, b) investigated in detail different face processing skills in a family of 13 members of four generations. Despite all family members having normal (or glasses corrected) vision, seven of them showed severe problems in recognizing faces of family members. A detailed neuropsychological assessment of different face processing mechanisms such as face detection, holistic processing, familiar face recognition and face perception indicated a rather heterogeneous cognitive phenotype both in adults and children. This not only confirms that congenital prosopagnosia represents an heterogeneous condition where sub-processes of face processing can be selectively impaired, but also indicated that this heterogeneity is within a single family that supposedly share a common genetic factor. To conclude, these and other studies support the key role played by our genes in determining how good we are in face recognition.

---

[10] It should be noted however that both parents in this family were affected by CP, thus the results may be accounted for by other inheritance patterns as well.

3.2.8 Neural Features of Congenital Prosopagnosia

The Haxby neural model of face processing indicates that different brain regions operate in concert to give rise to normal face processing skills. Amongst the regions crucially involved in face recognition, the FFA is the one that received much attention in neuroimaging studies of congenital prosopagnosia (Fig. 3.5).

The reason is that the FFA is the most "reliable" face area, that is, it is easily identifiable in virtually all people with normal face recognition skills. The most natural hypothesis most researchers had before starting experiments in congenital prosopagnosia was that the FFA should show abnormal functioning in this population. This is what has been found in the first single-case studies (Bentin et al. 2007; Hadjikhani and De Gelder 2002). However, albeit the first single case investigations pointed towards aberrant activity of the FFA, more recent studies conducted on small samples of CPs demonstrated that a typical position, dimension and intensity of the signal from the FFA was detected when both static and dynamic images of faces were presented (Avidan and Behrmann 2009; Avidan et al. 2005).

These results, somewhat counter intuitively, lead to the hypothesis that the FFA, despite being detectable in CP, could only be sensitive to the vision of faces (i.e., neurons spike in the presence of faces within the visual fields), but it cannot

Fig. 3.5 The FFA is anatomically localized within the brain by using 'localizer scan', in which blocks of faces and other objects are presented (e.g., 20 faces, 20 objects, 20 bodies, etc.). Due to space reason, only four items for categories are indicated in the figure (thanks to Dr Schmalzl for providing these pictures)

distinguish between different identities. To test this hypothesis a 'repetition-suppression task' was used, which causes a reduction of the BOLD signal when faces are repeated (Fig. 3.6).

In healthy controls, the repetition of a face induces a decrease of the neural activity because the FFA 'recognizes the face which has been repeated'; the decrement of the signal does not occur when two different faces are repeated in a sequence.

Results obtained by the repetition task demonstrated that the FFA in congenital prosopagnosia is not only sensitive to the category "face", but it is also able to discriminate among different identities (Kanwisher and Yovel 2006). Taken together, these results indicate that the FFA is necessary but not sufficient for typical face processing. In line with this idea, it has been demonstrated that, despite normal fusiform gyrus functioning, the activity of the anterior temporal lobe was less sensible, compared to controls, in representing a visual object as a face. This, in agreement with studies conducted in the patient population (Evans et al. 1995; Gainotti 2007; Williams et al. 2006), suggests that the anterior temporal lobe could also play a crucial role as a neurophysiological substrate of congenital prosopagnosia (Avidan 2012; Rivolta et al. 2011). Research on the issue is still in the early stages and hopefully future research will clarify the link between behaviour and neural activity in CP.

It should be noted that the results from functional neuroimaging studies reported above are very heterogeneous. This could be due to the different experimental designs adopted and/or the behavioural heterogeneity of people suffering from CP. In fact, only when larger sample sizes will be available, will neuroimaging studies be more accurate and able to divide the participants in subgroups like apperceptive and associative prosopagnosics.

Anatomical neuroimaging studies have, so far, reported the clearest results. We know that the brains of people with congenital prosopagnosia shows a reduced volume of the inferior temporal lobe, including the fusiform gyrus, and more anterior

Fig. 3.6 Repetition task. Couples of faces are presented in rapid sequence in two conditions: repeated and non-repeated stimuli. The repetition causes a decrease of the BOLD signal, called a 'neural adaptation'

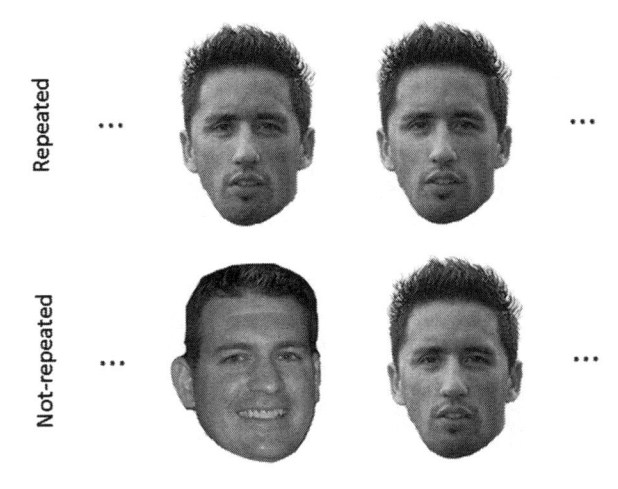

Fig. 3.7 Simplified
"neuroanatomical model" of
CP. Face-selective areas such
as the Occipital Face Area
(OFA), Fusiform Face Area
(FFA) and Anterior Temporal
cortex (AT) are connected
via the inferior longitudinal
fasciculus (shown in *green*).
Picture taken from Rivolta et
al. (2013), *Neuropsychology
Review*

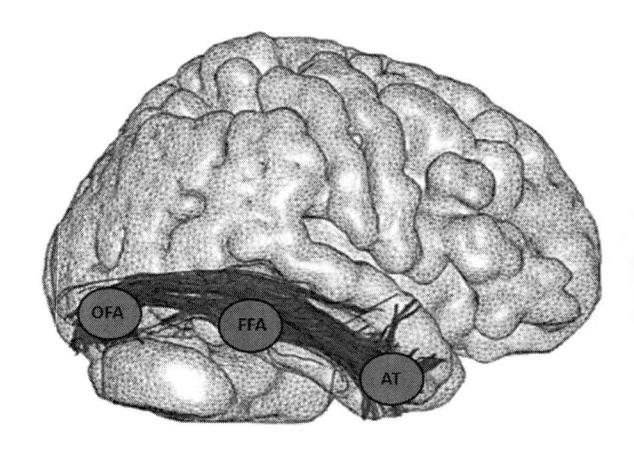

temporal areas (Behrmann et al. 2007; Garrido et al. 2009). Moreover, DTI analysis reported reduced volume of white matter fibres, such as the inferior longitudinal fasciculus, which connect posterior (i.e., occipital) to anterior (i.e., anterior temporal) face-sensitive regions in the brain of congenital prosopagnosics (Thomas et al. 2009). Analysis of structural connectivity also showed that the reduction of fibre tracts is selective; it is not so general as to involve all the white matter or other fasciculi, such as the corpus callosum (Fig. 3.7). To conclude, structural neuroimaging results demonstrate that the brain of people suffering from congenital prosopagnosia is structurally abnormal, due to changes in both white and grey matter.

Insights into the neural substrate of congenital prosopagnosia can also come from research in other developmental disorders, such as developmental dyslexia. It has been proposed that developmental dyslexia can be caused by a neural migration error. This means that, before the birth, clusters of circa 50–100 neurons may end their migration in wrong layer of the cortex, thus leading to cortical abnormalities/disorganization. The locus of this migration error would, in turn, determine the main manifested cognitive deficit. Even though there is no direct evidence, we can speculate that neural migrations involving the occipito-temporal regions may lead to congenital prosopagnosia (Dalrymple et al. 2013; Ramus 2004).

3.2.9 Neurophysiological Correlates

ERPs and MEG studies indicated the existence of face-specific components. The N/M170 represents by far the most investigated of them (Bentin et al. 1996). This component, which originates from the fusiform gyrus and the lateral occipital lobe (Rivolta et al. 2012) is generated by the activity of thousands of neurons that synchronously fire when a faces is within the visual field. Early experiments conducted in subjects with acquired prosopagnosia demonstrated that the N/M170, unlike controls, was not face-specific (i.e., other visual stimuli, such as watches, generated an N/M170 with the same amplitude as the one generated by faces), thus positing for a key role of this component for normal face processing. As such, the N/M170 in prosopagnosic people was the object of many studies with the aim to better clarify its features. Out of fourteen congenital prosopagnosia single-case studies published up to now, five had a normal N/M170, with higher amplitude for faces than for other categories, whereas nine cases reported an abnormal N/M170, where both faces and objects generated a similar component (Bentin et al. 2007; Bentin et al. 1999; Harris et al. 2005; Kress and Daum 2003; Minnebusch et al. 2007; Righart and de Gelder 2007). Is it very hard to understand the reason of the M170 variability in single cases because no correlation between the M170 amplitude and behaviour (i.e., face memory or face recognition skills) has ever been reported. In other words, there was not a clear indicator of why the M170 should present normal features in one subject and not in another.

Recently, our research group at MACCS completed an MEG study where we recorded the neural activity of face processing in CPs. Results, in line with previous studies, showed that the M170 is a 'dual' component that originates both from the occipital and the temporal lobe. Furthermore, as a group, CPs showed a face-sensitive M170 with similar features seen in control subjects (Fig. 3.8).

This has been recently confirmed in a group of 16 people with congenital prosopagnosia (Towler et al. 2012). Most importantly, thanks to the correlation between behavioural and neural data in the prosopagnosic population we found that the occipital M170 encodes for holistic mechanisms, while the fusiform M170 is responsible for the analysis of face features. Given that previous research in congenital prosopagnosia focuses on sensor activity, ignoring source activity, the heterogeneity of findings could be due to the lack of sensitivity of previous recordings, that detected only a single N/M170.

3.2.10 Can We "Cure" Prosopagnosia?

The study of anatomical and behavioural correlates of congenital prosopagnosia is not only important to better understand the mechanisms necessary for the normal face processing, but it also sheds light on the development of specific 'treatments'. This is what a person with congenital prosopagnosia reported:

> [...] The time may well be ripe for a prosopagnosia support group. And it may also be time to make sure that prosopagnosiacs [sic] are entitled to the same rights as other disabled people: it would make sense, for example, to require nametags (above all legibly lettered ones!) at all federally funded public gatherings. Most important of all, we should all work toward increased research funding that may eventually enable successful surgical or pharmacological interventions for those with an improperly functioning fusiform gyrus. Passivity is no longer a viable option: Prosopagnosiacs [sic] of the world, unite! (from Behrmann et al. 2009, pp. 189–190).

This person and many other people affected by this disabling condition are hoping that research could have more funds to study a possible cure prosopagnosia.

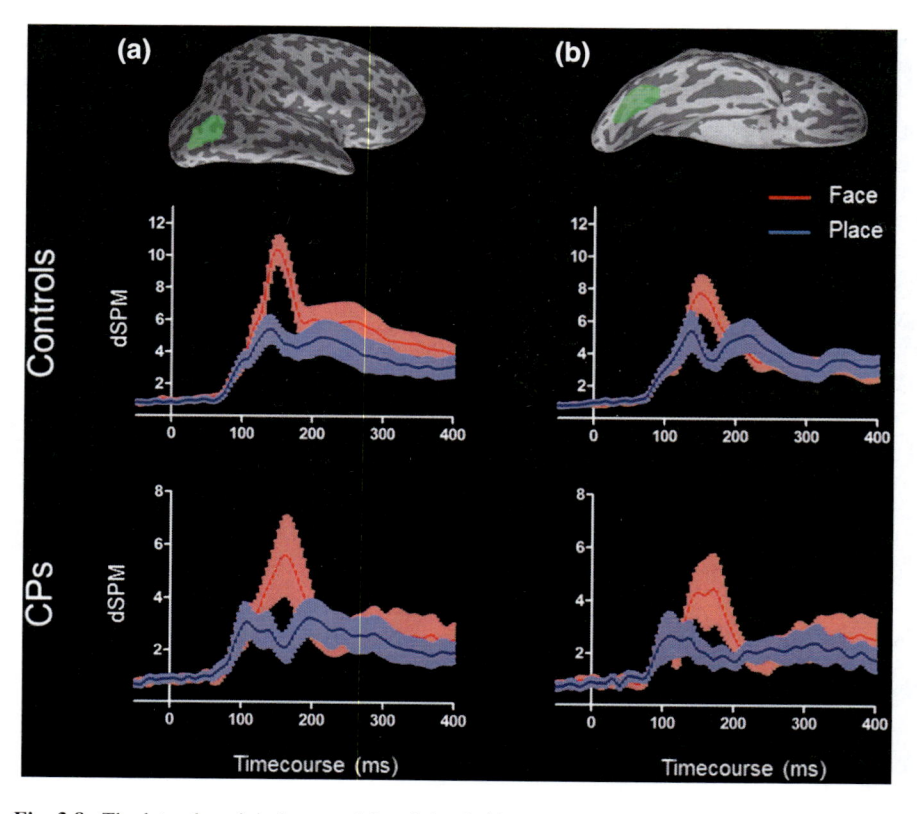

Fig. 3.8 The lateral occipital cortex (**a**) and the fusiform gyrus (**b**) areas that generate the M170 are shown in *green* on a brain surface. The activity for faces (*red*) is higher than the activity for places (*blue*) in both areas and in both groups. Picture taken with permissions from Rivolta et al. (2012), *Frontiers in Human Neuroscience*

This hope is legitimate if we think about the efficient improvements obtained by cognitive therapy in the disorders of reading (dyslexia), calculation (dyscalculia) or language (aphasia). In other words, why do we have the 'logopedist' but not the prosopedists?[11] Considering that, worldwide, millions of people could be affected by prosopagnosia, it is curious that there is still no answer to this question.

Some of you might wonder: Is it possible to become 'normal' at recognising faces? What are the techniques? Are drugs needed? So far, there have been no drugs developed to treat prosopagnosia, there is no surgical intervention to improve the functioning of brain areas and there will probably not be any of these in near future. In fact, only a richer knowledge about neuronal, genetic and molecular underpinning of face recognition could, potentially, lead to drug discovery.

---

[11] The term *prosopedist* does not exist in the current research literature. It has been suggested by me and inserted here for the first time to indicate the figure of a specialist with adequate skills to help people with prosopagnosia.

Encouraging results recently demonstrated that in normal subjects, the neuropeptide 'oxytocin' improves the performance of tasks investigating the memory of faces (Savaskan et al. 2008). The efficacy of oxytocin to improve the skills of face recognition in CP has recently been demonsrated in a sample of 10 people with congenital prosopagnosia (Bate et al., in press) (see, however, some skepticism in Herzmann et al. 2013). However, I personally think that drug discovery should not represent the only way to tackle prosopagnosia. In fact, most of the evidence of effective intervention is based on behavioural, neuropsychological, interventions.

Few studies over the last 6 years focused on the pattern of eye movements during face scanning. By monitoring the ocular movements of healthy people, scientists understood that the region around the eyes is the one with the largest number of fixations. This pattern is often not reflected by prosopagnosics, who rather focus on external features such as hairstyle, ears and forehead. Two single case studies conducted on an eight and a 4 years old kids with congenital prosopagnosia demonstrated that the training focused on scanning the eye region of the face improves their ability to recognize it (Brunsdon et al. 2006; Schmalzl et al. 2008b).

The improvement observed was not specific only for the images used in the training sessions but it also generalized to other faces. Therefore, extensive training aimed to reprogram the ocular movements might represent a valid and efficient rehabilitation strategy to overcome, or at least ameliorate, face recognition difficulties in congenital prosopagnosia. It is also important to note that the positive effects of oxytocin are associated with an increase of ocular fixations around the region of the eyes in healthy people (Guastella et al. 2008).

A different intervention towards the improving of face skills in prosopagnosia specifically focused on holistic processing. In a single case study, an adult with congenital prosopagnosia (MZ) was trained to improve the detection of differences in the spaces between different face features. MZ completed a battery of face recognition tasks before and after the training. His face recognition skills were significantly better after the training; this was true both for specific face processing tasks completed in the lab and for everyday life. Moreover, it was possible to identify the neurophysiological correlate behind this improvement: The N170 became face-specific and the functional connectivity between OFA and FFA increased only after the treatment.

These are, so far, the most relevant studies which investigated potential ways to ameliorate face recognition skills in people with congenital prosopagnosia. It will be important for future research to conduct experiments on a larger scale of individuals with prosopagnosia, and to monitor the changes over time of neural and behavioural patterns; for instance, before, soon after and several months/years after rehabilitation. This direction, in my view, should have maximum priority as people with congenital prosopagnosia not only report fastidious difficulties in recognising faces, but they also often refer psychological distress such as anxiety, feelings of inadequacy and social phobia. This contributes to their difficulty in making and maintaining friendship, complications at work/school, and, in children, risk of being put in dangerous situations with strangers (Dalrymple et al. 2013; Diaz 2008; Fine 2013; Yardley et al. 2008).

3.3 Conclusions

In this chapter we learned that prosopagnosia exists in two different forms: acquired and congenital. Albeit the acquired form was the first to be described in the scientific community, congenital prosopagnosia, affecting 2–2.9 % of the general population, represents the most popular form of prosopagnosia. This is the reason why I decided to mainly focus the book on the congenital form. We also learned that congenital prosopagnosia is very heterogeneous at a behavioural level, in which people might show a different 'cognitive phenotype', even when they belong to the same family, suggesting that we could therefore subcategorize prosopagnosia in apperceptive and associative. We also know that there is a genetic component involved in the condition and that the key genes are still a matter of scientific investigation. Finally, we know that there are possible behavioural strategies that can improve the ability to recognize faces and that, in the future, drugs may be developed to help the work of clinicians. What we do not know so far is at which extent people with congenital prosopagnosia can still recognise faces without being aware of it. This form of implicit recognition, called 'covert recognition', has theoretical and clinical importance and it will represent the topic of the next chapter.

References

Avidan, G. (2012). *Selective dissociation between core and extended regions of the face processing network in congenital prosopagnosia* Paper presented at the Annual meeting of the Society for Neuroscience, New Orleans.

Avidan, G., & Behrmann, M. (2009). Functional MRI reveals compromised neural integrity of the face processing network in congenital prosopagnosia. *Current Biology, 19*, 1–5.

Avidan, G., Hasson, U., Malach, R. l, and Behrmann, M. (2005). Detailed exploration of face-related processing in congenital prosopagnosia: 2. functional neuroimaging findings. *Journal of Cognitive Neuroscience, 17*, 1150–1167.

Baron-Cohen, S., Wheelwright, S., Skinner, R., Martin, J., & Clubley, E. (2001). The autism-spectrum quotient (AQ): Evidence from Asperger syndrome/high-functioning autism, males and females, scientists and mathematicians. *Journal of Autism and Developmental Disorders, 31*, 5–17.

Barton, J. J. (2008a). Structure and function in acquired prosopagnosia: Lesson from a series of 10 patients with brain damage. *Journal of Neuropsychology, 2*, 197–225.

Barton, J. J. (2008b). Structure and function in acquired prosopagnosia: Lessons from a series of 10 patients with brain damage. *Journal of Neuropsychology, 2*, 197–225.

Barton, J. J., Cherkasova, Mariya V., Press, Daniel Z., Intriligator, James M., & O'Connor, Margaret. (2003). Developmental prosopagnosia: A study of three patients. *Brain and Cognition, 51*, 12–30.

Bate, S., Cook, S. J., Duchaine, B., Tree, J. J., Burns, E. J., Hodgson, T. L. *Intranasal Inhalation of Oxytocin Improves Face Processing in Developmental Prosopagnosia*, Cortex. (in press).

Behrmann, M., & Avidan, G. (2005). Congenital prosopagnosia: Face-blind from birth *Trends in Cognitive Neuroscience, 9*, 180–187.

Behrmann, M., Avidan, G., Marotta, J. J., & Kimchi, R. (2005). Detailed Exploration of Face-related Processing in Congenital Prosopagnosia: 1 Behavioral Findings. *Journal of Cognitive Neuroscience, 17*, 1130–1149.

Behrmann, M., Avidan, G., Gao, F., & Black, S. (2007). Structural imaging reveals anatomical alterations in inferotemporal cortex in congenital prosopagnosia. *Cerebral Cortex, 17*, 2354–2363.

Behrmann, M., Avidan, G., Thomas, C., & Humphreys, K. (2009). Congenital and acquired prosopagnosia: Flip sides of the same coin? In I. Gauthier, M. Tarr, & D. Bub (Eds.), *Perceptual expertise: Bridging brain and behaviour.* Oxford: Oxford University Press.

Bentin, S., McCarthy, G., Perez, E., Puce, A., & Allison, T. (1996). Electrophysiological studies of face perception in humans. *Journal of Cognitive Neuroscience, 8*, 551–565.

Bentin, S., Deouell Leon, Y., and Soroker, N. (1999). Selective visual streaming in face recognition: Evidence from developmental prosopagnosia. *Neuroreport, 10*, 823–827.

Bentin, S., DeGutis, Joseph M., D'Esposito, Mark, & Robertson, Lynn C. (2007). Too many trees to see the forest: Performance, event-related potential, and functional magnetic resonance imaging manifestations of integrative congenital prosopagnosia. *Journal of Cognitive Neuroscience, 19*, 132–146.

Benton, A. L. (1990). Facial recognition. *Cortex, 26*, 491–499.

Bodamer, J. (1947). Die Prosop-agnosie. *Archiv fur Psychiatrie und Nervkrankheiten, 179*, 6–53.

Bowles, D. C., McKone, E., Dawel, A., Duchaine, B., Palermo, R., Schmalzl, L., Rivolta, D., Wilson, E. C., and Yovel, G. (2009). Diagnosing prosopagnosia: Effects of aging, sex, and participant-stimulus ethnic match on the Cambridge face memory test and Cambridge face perception test *Cognitive Neuropsychology, 26*, 423–455.

Braun, C. M., Denault, C., Cohen, H., & Rouleau, I. (1994). Discrimination of facial identity and facial affect in temporal and frontal lobectomy patients. *Brain and Cognition, 24*, 198–212.

Brunsdon, R., Coltheart, M., Nickels, L., & Joy, P. (2006). Developmental prosopagnosia: A case analysis and treatment study. *Cognitive Neuropsychology, 23*, 822–840.

Busigny, T., Joubert, S., Felician, O., Ceccaldi, M., & Rossion, B. (2010). Holistic perception of the individual face is specific and necessary: Evidence from an extensive case study of acquired prosopagnosia. *Neuropsycholgia, 48*, 4057–4092.

Dalrymple, K. A., Corrow, S., Yonas, A., and Duchaine, B. (2013). Developmental prosopagnosia in childhood. *Cognitive Neuropsychology, iFirst*, 1–26.

Damasio, A. R., Tranel, D., & Damasio, H. (1990). Face agnosia and the neural substrates of memory. *Annual Reviews Neuroscience, 13*, 89–109.

De Haan, E. H. F., & Campbell, R. A. (1991). Fifteen year follow-up of a case of developmental prosopagnosia.Cortex, *27*(4): 489–509.

De Haan, E. (1999). A familial factor in the development of face recognition deficits. *Journal of Clinical and Experimental Neuropsychology, 21*, 312–315.

De Haan, E. (2001). Face perception and recognition. In B. Rapp (Ed.), *The handbook of cognitive neuropsychology—What deficits reveal about the human mind.* Philadelphia: Psychology Press.

Della, S. S., and Young, A. W. (2003). Quaglino's 1867 case of prosopagnosia. *Cortex, 39*, 533–540.

Denes, G., & Pizzamiglio, L. (1996). *Manuale di neuropsicologia. Normalita' e patologia dei processi cognitivi* (2nd ed.). Bologna: Zanichelli.

Dennett, H. W., McKone, E., Tavashmi, R., Hall, A., Pidcock, M., Edwards, M., and Duchaine, B. (2011). The Cambridge car memory test: A task matched in format to the Cambridge face memory test, with norms, reliability, sex differences, dissociations from face memory, and expertise effects. *Behavior Research Methods.*

Diaz, A. L. (2008). Do i know you? A case study of prosopagnosia. *The Journal of School Nursing, 24*, 284–289.

Duchaine, B. (2011). Developmental prosopagnosia: Cognitive, neural and developmental investigations. In A. J. Calder, G. Rhodes, M. Johnson, & J. V. Haxby (Eds.), *The oxford handbook of face perception.* Oxford: Oxford University Press.

Duchaine, B., & Nakayama, K. (2005). Dissociations of face and object recognition in developmental prosopagnosia. *Journal of Cognitive Neuroscience, 17*, 249–261.

Duchaine, B., & Nakayama, K. (2006a). The Cambridge face memory test: Results for neurologically intact individuals and an investigation of its validity using inverted face stimuli and prosopagnosic participants. *Neuropsychologia, 44*, 576–585.

Duchaine, B., & Nakayama, K. (2006b). Developmental prosopagnosia: A window to content-specific face processing. *Current Opinion in Neurobiology, 16*, 166–173.

Duchaine, B., Parker, H., and Nakayama, K. (2003). Normal recognition of emotion in a prosopagnosic. *Perception, 32*, 827–838.

Duchaine, B., Yovel, G., Butterworth, E. J., and Nakayama, K. (2006). Prosopagnosia as an impairment to face-specific mechanisms: Elimination of the alternative hypotheses in a developmental case. *Cognitive Neuropsychology, 23*, 714–747.

Duchaine, B., Germine, L., & Nakayama, K. (2007). Family resemblance: Ten family members with prosopagnosia and within-class object agnosia. *Cognitive Neuropsychology, 24*, 419–430.

Evans, J. J., Heggs, A. J., Antoun, N., & Hodges, J. R. (1995). Progressive prosopagnosia associated with selective right temporal lobe atrophy: A new syndrome? *Brain, 118*, 1–13.

Ewbank, M. P., Henson, R. N., Rowe, J. B., Stoyanova, R. S., & Calder, A. J. (2012). Different neural mechanisms within occipitotemporal cortex underlie repetition suppression across same and different-size faces. *Cerebral Cortex*. doi:10.1093/cercor/bhs1070.

Farah, M. J. (1996). Is face recognition 'special'? Evidence from neuropsychology. *Behavioural Brain Research, 76*, 181–189.

Farah, M. J., Tanaka, J. W., & Drain, H. M. (1995). What causes the face inversion effect? *Journal of Experimental Psychology: Human Perception and Performance, 21*, 628–634.

Farah, M. J., Rabinowitz, C., Quinn, G. E., & Liu, G. (2000). Early commitment of neural substrates for face recognition. *Cognitive Neuropsychology, 17*, 117–123.

Fine, D. R. (2013). A life with prosopagnosia. *Cognitive Neuropsychology, iFirst*, 1–6.

Gainotti, G. (2007). Face familiarity feelings, the right temporal lobe and the possible underlying neural mechanisms. *Brain Research Reviews, 56*, 214–235.

Gainotti, G., & Marra, C. (2011). Differential contribution of right and left temporo-occipital and anterior temporal lesions to face recognition disorders. *Frontiers in Human Neuroscience, 5*, 1–11.

Garrido, L., Furl, N., Draganski, B., Weiskopf, N., Stevens, J., Chern-Yee, T. G., Driver, J., Dolan, R. J., and Duchaine, B. C. (2009). Voxel-based morphometry reveals reduced grey matter volume in the temporal cortex of developmental prosopagnosics. *Brain, 132*, 3443–3455.

Grueter, M., Grueter, T., Bell, V., Horst, J., Laskowski, W., Sperling, K., et al. (2007). Hereditary prosopagnosia: The first case series. *Cortex, 43*, 734–749.

Guastella, A. J., Mitchell, P. B., & Dadds, M. R. (2008). Oxytocin increases gaze to the eye region of human faces. *Biological Psychiatry, 63*, 3–5.

Hadjikhani, N., & De Gelder, B. (2002). Neural basis of prosopagnosia: An fMRI study. *Human Brain Mapping, 16*, 176–182.

Harris, A., Duchaine, B. C., and Nakayama, K. (2005). Normal and abnormal face selectivity of the M170 response in developmental prosopagnosics. *Neuropsychologia, 43*, 2125–2136.

Herzmann, G., Young, B., Bird, C. W., and Curran, T. (2013). Oxytocin can impair memory for social and non-social visual objects: A within-subject investigation of oxytocin's effects on human memory. *Brain Research. Cognitive Brain Research*.

Humphreys, K., Avidan, G., & Behrmann, M. (2007). A detailed investigation of facial expression processing in congenital prosopagnosia as compared to acquired prosopagnosia. *Experimental Brain Research, 176*, 356–373.

Jones, R. D., & Tranel, D. (2001). Severe developmental prosopagnosia in a child with superior intellect. *Journal of Clinical and Experimental Neuropsychology, 23*(3):265–273.

Kanwisher, N. (2010). Functional specificity in the human brain: A window into the functional architecture of the mind. *Proceedings of the National Academy of Science USA, 107*, 11163–11170.

Kanwisher, N., & Yovel, G. (2006). The fusiform face area: A cortical region specialized for the perception of faces. *Philosophical Transactions of the Royal Society B, 361*, 2109–2128.

Kempler, D. (2005). *Neurocognitive Disorders in Aging* (1st ed.). Thousand Oaks, California, US: Sage Publications, Inc.

Kennerknecht, I., Grueter, T., Welling, B., Wentzek, S., Horst, J., Edwards, S., et al. (2006). First Report of Prevalence of Non-Syndromic Hereditary Prosopagnosia (HPA). *American Journal of Medical Genetics Part A, 140A*, 1617–1622.

Kress, T., & Daum, I. (2003). Developmental prosopagnosia: A review. *Behavioural Neurology, 14*, 109–121.

Lang, N., Baudewig, J., Kallenberg, K., Antal, A., Happe, S., Dechent, P., et al. (2006). Transient prosopagnosia after ischemic stroke. *Neurology, 66*, 916.

Le Grand, R., Cooper, P. A., Mondloch, C. J., Lewis, T. L., Sagiv, N., de Gelder, B., and Maurer, D. (2006). What aspects of face processing are impaired in developmental prosopagnosia? *Brain and Cognition, 61*, 139–158.

Lee, Y., Duchaine, B. C., Wilson, H. R., and Nakayama, K. (2010). Three cases of developmental prosopagnosia from one family: Detailed neuropsychological and psychophysical investigation of face processing. *Cortex, 46*, 949–964.

Malach, R., Reppas, J. B., Benson, R. R., Kwong, K. K., Jiang, H., Kennedy, W., et al. (1995). Object-related activity revealed by functional magnetic resonance imaging in human occipital cortex. *Proceedings of the National Academy of Sciences (PNAS) US, 92*, 8135–8139.

Mattson, A. J., Levin, H. S., & Grafman, J. (2000). A case of prosopagnosia following moderate closed head injury with left hemisphere focal lesion. *Cortex, 36*, 125–137.

McConachie, H. R. (1976). Developmental prosopagnosia: A single case report. *Cortex, 12*, 76–82.

McKone, E., Hall, A., Pidcock, M., Palermo, R., Wilkinson, R. B., Rivolta, D., et al. (2011). Face ethnicity and measurement reliability affect face recognition performance in developmental prosopagnosia: Evidence from the Cambridge Face Memory Test–Australian. *Cognitive Neuropsychology, 28*, 109–146.

Meadows, J. C. (1974). The anatomical basis of prosopagnosia. *Journal of Neurology, Neurosurgery and Psychiatry, 37*, 489–501.

Minnebusch, D. A., Suchan, B., Ramon, M., & Daum, I. (2007). Event-related potentials reflect heterogeneity of developmental prosopagnosia. *European Journal of Neuroscience, 25*, 2234–2247.

Moscovitch, M., Winocur, G., & Behrmann, M. (1997). What is special about face recognition? Nineteen experiments on a person with visual objects agnosia and dyslexia but normal face recognition. *Journal of Cognitive Neuroscience, 9*, 555–604.

Palermo, R., Rivolta, D., Wilson, C. E., & Jeffery, L. (2011a). Adaptive face space cpding in congenital prosopagnosia: Typical figural aftereffects but abnormal identity aftereffects. *Neuropsychologia, 49*, 3801–3812.

Palermo, R., Willis, M. L., Rivolta, D., McKone, E., Wilson, C. E., & Calder, A. J. (2011b). Impaired holistic coding of facial expression and facial identity in congenital prosopagnosia. *Neuropsychologia, 49*, 1226–1235.

Quaglino, A., Borelli, G. B., Della Sala, S., & Young, A. W., (2003). *Quaglino's 1867 case of prosopagnosia.* Cortex, 39(3):533–540.

Ramus, F. (2004). Neurobiology of dyslexia: A reinterpretation of the data. *Trends in Neuroscience, 27*, 720–726.

Riddoch, M. J., Johnston, R. A., Bracewell, B. M., Boutsen, L., & Humphreys, G. W. (2008). Are faces special? A case of pure prosopagnosia. *Cognitive Neuropsychology, 25*, 3–26.

Righart, R., & de Gelder, B. (2007). Impaired face and body perception in developmental prosopagnosia. *Proceedings of the National Academy of Science USA, 104*, 17234–17238.

Rivolta, D. (2010). *Face processing in typical and congenitally prosopagnosic adults: Behavioural and neuroimaging investigations.* PhD: Macquarie University, Sydney.

Rivolta, D., Schmalzl, L., Coltheart, M., & Palermo, R. (2010). Semantic information can facilitate covert face recognition in congenital prosopagnosia. *Journal of Clinical and Experimental Neuropsychology, 32*, 1002–1016.

Rivolta, D., Schmalzl, L., Palermo, R., and Williams, M. A. (2011). *Multi-voxel pattern analysis of fMRI data reveals abnormal anterior temporal lobe activity in congenital prosopagnosia.* Paper presented at the European Conference on Visual Perception (ECVP) Toulouse.

Rivolta, D., Palermo, R., Schmalzl, L., & Williams, M. A. (2012). Investigating the features of the M170 in congenital prosopagnosia. *Frontiers in Human Neuroscience, 6*, 45.

Rossion, B. (2009). Clarifying the functional neuroanatomy of face perception by single case neuroimaging studies of acquired prosopagnosia. In Michael Jenkin & Laurence Harris (Eds.), *Cortical Mechanisms of Vision* (pp. 171–207). Cambridge: Cambridge University Press.

Rossion, B., Caldara, R., Seghier, M., Schuller, A.-M., Lazeyras, F., & Mayer, E. (2003). A network of occipito-temporal face-sensitive areas besides the right middle fusiform gyrus is necessary for normal face processing. *Brain, 126*, 2381–2395.

Savaskan, E., Ehrhardt, R., Schultz, A., Walter, M., & Schächinger, H. (2008). Post-learning intranasal oxytocin modulates human memory for facial identity. *Psychoneuroendocrinology, 33*, 368–374.

Schmalzl, L. (2007). *Fractionating face processing in congenital prosopagnosia*. Ph.D., Macquarie University, Sydney, Australia. (BF242,S48).

Schmalzl, L., Palermo, R., & Coltheart, M. (2008a). Cognitive heterogeneity in genetically based prosopagnosia: A family study. *Journal of Neuropsychology, 2*, 99–117.

Schmalzl, L., Palermo, R., Green, M., Brunsdon, R., & Coltheart, M. (2008b). Training of familiar face recognition and visual scan paths for faces in a child with congenital prosopagnosia. *Cognitive Neuropsychology, 25*, 704–729.

Schultz, R. T. (2005). Developmental deficits in social perception in autism: The role of the amygdala and fusiform face area. *International Journal of Developmental Neuroscience, 23*, 125–141.

Sergent, J., & Signoret, J. L. (1992). Varieties of functional deficits in prosopagnosia. *Cerebral Cortex, 2*, 375–388.

Sparr, S. A., Jay, M., Drislane, F. W., & Venna, N. (1991). A historic case of visual agnosia revisited after 40 years. *Brain, 114*, 789–800.

Steeves, J. K. E., Culham, J. C., Duchaine, B., Cavina Pratesi, C., Valyear, K., Schindler, I., Humphrey, K., Milner, A. D., and Goodale, M. A. (2006). The fusiform face area is not sufficient for face recognition: Evidence from a patient with dense prosopagnosia and no occipital face area. *Neuropsycholgia, 44*, 596–609.

Steeves, J. K. E., Dricot, L., Golzz, H. C., Sorger, B., Peters, J., Milner, A. D., et al. (2009). Abnormal face identity coding in the middle fusiform gyrus of two brain-damaged prosopagnosics patients. *Neuropsycholgia, 47*, 2584–2592.

Sugimoto, A., Miller, M. W., Kawai, Y., Shiota, J., & Kawamura, M. (2011). Another piece of the jigsaw: A case report of prosopagnosia with symptomatological, imaging and post mortem anatomical evidence. *Cortex.* doi:10.1016/j.cortex.2011.1004.1003.

Susilo, T., McKone, E., Dennett, H., Darke, H., Palermo, R., Hall, A., et al. (2011). Face recognition impairments despite normal holistic processing and face space coding: Evidence from a case of developmental prosopagnosia. *Cognitive Neuropsychology, 27*, 636–664.

Thomas, C., Avidan, G., Humphreys, K., Gao, F., and Behrmann, M. (2009). Reduced structural connectivity in ventral visual cortex in congenital prosopagnosia. *Nature Neuroscience, 12*, 29–31.

Towler, J., Gosling, A., Duchaine, B., and Eimer, M. (2012). The face-sensitive N170 component in developmental prosopagnosia. *Neuropsycholgia,* http://dx.doi.org/10.1016/j.neuropsychologia.2012.10.017.

Wada, Y., & Yamamoto, T. (2001). Selective impairment of facial recognition due to haematoma restricted to the right fusiform and lateral occipital area. *Journal of Neurology, Neurosurgery and Psychiatry, 71*, 254–257.

Warrington, E. K. (1984). *Recognition memory test*, Windsor (UK): NFER-NELSON.

Williams, Mark, Savage, Greg, & Halmagyi, Michael. (2006). Abnormal configural face perception in a patient with right anterior temporal lobe atrophy. *Neurocase, 12*, 286–291.

Williams, M. A., Berberovic, N., & Mattingley, J. B., (2007). Abnormal FMRI adaptation to unfamiliar faces in a case of developmental prosopamnesia. *Current Biology, 17*(14):1259–1264.

Wilson, C. E., Freeman, P., Brock, J., Burton, A. M., & Palermo, R. (2010a). Facial identity recognition in the broader autism phenotype. *PLoS ONE, 5*, e12876. doi:10.1371/journal.pone.0012876.

Wilson, C. E., Palermo, R., Schmalzl, L., and Brock, J. (2010b). Specificity of impaired facial identity recognition in children with suspected developmental prosopagnosia. *Cognitive Neuropsychology, 27*, 30–45.

Yardley, L., McDermott, L., Pisarski, S., Duchaine, B., & Nakayama, K. (2008). Psychosocial consequences of developmental prosopagnosia: A problem of recognition. *Journal of Psychosomatic Research, 65*, 445–451.

Chapter 4
Can I Recognize Faces Without Knowing it? Evidence of Covert Face Recognition in Prosopagnosia

4.1 Covert Recognition: A General Description

A brain lesion can often cause a dysfunction of a specific cognitive domain. For example, some brain lesions can leave individuals unable to remember old and new events of their lives, such as what they did the night before, where they completed their studies or their appointment with the dentist scheduled for next week. These individuals have developed a disorder known as "amnesia". Other individuals may develop what is known as "neglect" (or unilateral spatial negligence). These individuals have lost the ability to attend to (i.e., explicitly recognize) half of their visual space, despite normal vision.[1] For instance, individuals suffering from "neglect" will not shave the left part of their face, they will only eat food on the left side of the dish and they forget to put their left leg in their pants. As we know by now, other people develop "prosopagnosia" (Kempler 2005).

Might there be instances of covert or implicit recognition in any of the above cases? Covert or implicit recognition refers to the ability to process information, despite lack of awareness. This ability to unconsciously recognize information has been found in many domains, starting more than a century ago. Emblematic of this ability is the amnesic case described at the beginning of the twentieth century by Cleparede, a French neurologist. Even though they met many times, the amnesic patient was approaching the doctor as it was the first time they met. The doctor decided to hide a pin in his hand to see whether the patient could remember this slightly painful event each subsequent time they met. It appeared that, despite having no recollection of meeting the doctor previously, he become very reluctant to shake his hand, where the pin was usually hidden. In other words, the patient could not consciously access information from memory (i.e., he believed he never saw the doctor before), but could unconsciously use it (i.e., not to shake the doctor's hand).

---

[1] Very often the lesions causing neglect occur in the right hemisphere and therefore their symptoms often include lack of attention to the left side of their visual field

D. Rivolta, *Prosopagnosia*, Cognitive Systems Monographs 20,
DOI: 10.1007/978-3-642-40784-0_4, © Springer-Verlag Berlin Heidelberg 2014

Another famous case is described by Marshall and Halligan (1988), where they presented line drawings of houses to a patient with neglect. Some houses were normal, whereas other had smoke exiting from a window on the left-side (the neglected side). As expected the patient judged the houses to be identical. However, when forced to choose which house he would prefer to live in, he chose the house without smoke 17 out 21 times. This is much more often than predicted by chance, suggesting that he was able to unconsciously process the "dangerous" information present in picture.

Covert recognition has been also described in acquired prosopagnosia using behavioural tasks, electrophysiological techniques and autonomic measures. Overall, the study of these patients demonstrated that, despite the absence of awareness, they could recognize familiar faces, suggesting that the face recognition system was damaged by a brain lesion, but this lesion "blocked" only the explicit (naming a face), not the implicit ability do recognize faces. Recent investigations have also reported covert face recognition in congenital prosopagnosia. This is quite surprising considering that, contrary to people with acquired prosopagnosia, people with congenital prosopagnosia have never acquired normal representations of familiar faces. How can there be covert recognition of faces never recognized in the first place? Below we discuss why this may be the case and the main techniques adopted for investigating covert face recognition.

4.2 Covert Face Recognition in Congenital Prosopagnosia

People with congenital prosopagnosia cannot identify others by their faces. This means they have problems in overt or explicit face recognition; that is, they cannot name a face or provide any biographic information or whether it is familiar or not. Are, however, these faces recognized implicitly, without awareness? How can we detect implicit face recognition? What is covert face recognition in congenital prosopagnosia telling us? In examining covert face recognition, researchers have used a number of techniques including behavioural, electrophysiological and autonomic measures (for a review, see Rivolta et al. 2013; Schweinberger and Burton 2003).

4.2.1 Behavioural Techniques

Many behavioural tasks adopted in experimental psychology have been adapted to assess covert face recognition. Some examples are the Forced choice task, the Matching task and the Priming task.[2] In the Forced choice task, people are forced to select which of the two presented faces is the familiar one ("Forced choice familiarity

---

[2] The list of behavioural tasks that can be adopted for the investigation of covert face recognition is longer and includes tasks like the "Face-name interference task", the "Sorting task" and the "Learning task". These have only been adopted in acquired prosopagnosia and, as such, are not reported here. The interested reader should refer, for instance, to Schweinberger and Burton (2003).

Fig. 4.1   Forced choice tasks. **a** Forced choice familiarity task (here the famous face is on the left and belongs to Sylvester Stallone) and **b** Forced choice cued task (here Tom Cruise is on the *left* whereas Russel Crowe is on the *right*)

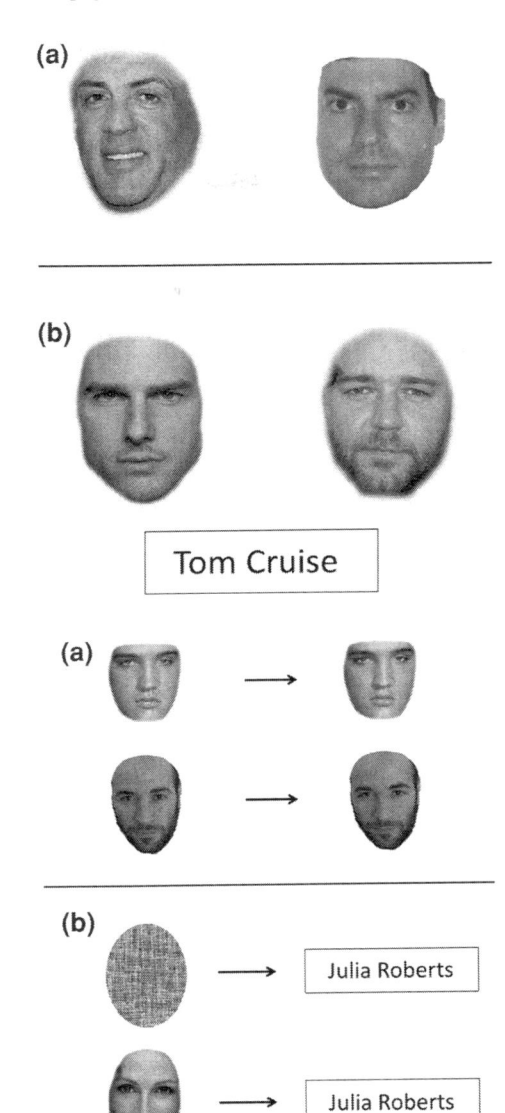

Fig. 4.2   In the Matching task **a** people have to tell whether two repeated faces belong to the same person or not. The faces could be familiar or unfamiliar. In the Priming task **b** people have to say whether a name belongs to an actor or politician. Names are shown after a scrambled face or after the actual face

task", see Fig. 4.1a) or which of the two faces shown best matches the presented name given as a cue ("Forced choice cued task", see Fig. 4.1b). In the "Matching task" people have to say whether the two faces shown in a sequence are the same or different (see Fig. 4.2a) and, in one version of the "Priming task", participants have to judge whether the presented name belongs to an actor or a politician, in the case when faces and names belong to the same person and when they do not (see Fig. 4.2b).

People without face recognition problems can, of course, perform perfectly on both the Forced choice tasks; they can easily tell familiar and unfamiliar faces apart, and they can easily match the name to the correct face. On the Matching task, results have indicated that controls are faster to match familiar than unfamiliar faces and, in the Priming task, they are able to categorize the presented name faster when the correct face precedes it.

Can the same results be found in individuals with congenital prosopagnosia, where people cannot identify faces overtly? The same effects found in this population would tell us that faces, albeit not explicitly recognized, can be (implicitly) processed by the visual system. Recent evidence demonstrated that, as a group, people with congenital prosopagnosia can discriminate familiar and unfamiliar faces more often than chance in the Forced choice familiarity task, despite their claims that they are merely guessing. Similarly, in the Forced choice cued task, they can match the name to the correct face more often than chance despite their inability to identify any of these faces explicitly. Again, people with congenital prosopagnosia are faster to match familiar than unfamiliar faces even though all faces look unfamiliar to them. Results on the priming task indicated, however, that the category of the face (actor vs. politician) does not influence the speed at which they categorize the name. This suggests that some tasks are more sensitive than others in detecting covert aspects of face recognition (Avidan and Behrmann 2008; Rivolta et al. 2012a; Rivolta et al. 2010).

4.2.2 Autonomic Response

Skin Conductance Response (SCR) has also been adopted in assessing whether individuals with prosopagnosia can covertly recognize faces. The first documentation of covert face recognition using SCR comes from a physiological experiment on an acquired case conducted by Bauer (1984). Bauer presented a series of photographs of familiar and unfamiliar faces to the patient. Even though the patient could not recognize any of faces overtly, his SCRs were larger when viewing a familiar compared to unfamiliar faces. As this pattern of activity (SCRs bigger for familiar with respect to unfamiliar) is routinely found in normal subjects, it has provided first physiological index of covert recognition. The main anatomical structure mediating this form of covert recognition is small and located deep in the brain; it is called the "amygdala". Researchers have found physiological indicators of covert recognition using SCR in children as young as 5 (Jones and Tranel 2001) and adults as old as 61 (Bate and Cook 2012) with congenital prosopagnosia.

4.2.3 Electrophysiological Techniques

Event Related Potentials (ERP) have also been successfully adopted in examining covert face processing (Renault et al. 1989). A recent ERP study demonstrated that people without face recognition problems show a negative deflection called the

N250, which is larger for familiar faces compared to unfamiliar faces (Gosling and Eimer 2011). Subsequently (Eimer et al., 2012), they found that six out of twelve individuals with congenital prosopagnosia exhibited the M250 for familiar faces, thereby demonstrating covert recognition.

4.3 The Relation Between Different Forms of Covert Face Recognition

Overall, the studies reported above indicate that face processing skills in congenital prosopagnosia are not completely damaged and that some form of recognition takes place and can be demonstrated using specific techniques. One current hypothesis suggests that there might be "degraded" face representations unable to trigger explicit face recognition, but that are good enough to mediate covert face processing. How can there be face representations good enough to mediate covert recognition, but too poor to support overt recognition? We can speculate that the formation of these degraded face representations may stem from a reduced number of saccades within the eyes-mouth-nose region (Schmalzl et al. 2008) and/or weak holistic processing (Avidan et al. 2011; Palermo et al. 2011).

As there's a correlation between overt and behavioural covert face recognition, that is, the more explicit face recognition is impaired, the more unlikely is to show behavioural covert face recognition (Barton et al. 2004; Rivolta et al. 2013; Schweinberger and Burton 2003), it's possible that there exists a single neuro-cognitive system that mediates both forms of face recognition. When this system is degraded (i.e., damaged in acquired prosopagnosia or underdeveloped in congenital prosopagnosia) it can still mediate covert face recognition, whereas when it is totally compromised it fails to mediate both covert and overt recognition (Rivolta et al. 2013). Conversely, physiological (i.e., as measured with SCR) covert face recognition seems to be mediated by a separated system from the one involved in overt recognition. Evidence of this comes from what neuropsychologists call "double dissociations" between people with prosopagnosia and patients with Capgras syndrome (people that believe that relatives have been substituted by aliens or impostors). These two are complementary cases: people with a cognitive problem recognizing faces (prosopagnosics) show intact SCR, whereas people without cognitive problems recognizing faces (Capgras patients) show abnormal SCR. This suggests that the anatomical and cognitive system that mediates overt face recognition is different from the system that mediates physiological responses of covert face recognition: they are somehow separated and can be selectively impaired (Fig. 4.3).

Up to now, unfortunately, there is still no comprehensive study investigating behavioural and physiological covert face recognition in the same sample of people with congenital prosopagnosia. In other words, we still do not completely know the relation between different kinds of covert recognition.

To conclude, the research on covert face recognition in congenital prosopagnosia is not only important for the theoretical point of view outlined above, but also

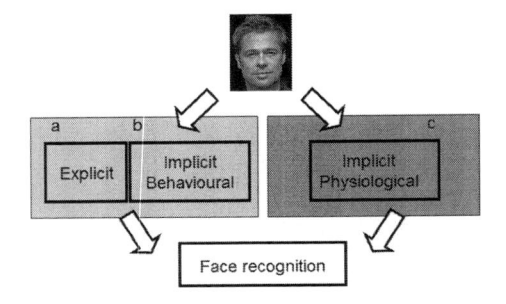

Fig. 4.3 Schematic representation of the link between *a* explicit, *b* implicit-behavioural and *c* implicit-physiological covert face recognition. Cases of prosopagnosia (both acquired and congenital) have shown that damage to "*a*" can be so serious to also compromise "*b*", or it may be mild, thereby permitting "*b*" to emerge. As such, "*a*" and "*b*" are mediated by the same system (green). However, irrespective of the entity of the damage in "*a*", "*c*" may be present because it is mediate by a different system (red). Conversely, patients with Capgras syndrome have no problems in "*a*" and "*b*", but show an impairment in "*c*"

suggest that possible strategies of intervention may be feasible. The message seems clear: face information is present in individuals with congenital prosopagnosia, albeit "trapped" within the visual system, unable to emerge. Clearly, more research is needed, but the existence of covert face recognition in individuals with congenital prosopagnosia provides hope that one day a successful intervention will be found.

4.4 A Neuroanatomical Model of Congenital Prosopagnosia

Can we integrate the main behavioural and neuroimaging findings described so far and speculate about a neuroanatomical model of congenital prosopagnosia? As should be clear by now, much research has demonstrated that a network of cortical and subcortical brain regions mediates human face processing. Within these regions, the lateral occipital lobe (Gauthier et al. 2000) and the fusiform gyrus (Kanwisher et al. 1997) represent "core regions" for normal face processing (Haxby et al. 2000). Previous investigations have indicated that, despite the clear behavioural impairment in face recognition, people with congenital prosopagnosia exhibit normal structure (Behrmann et al. 2007) and functioning (Avidan and Behrmann 2009; Avidan et al. 2005) of these core regions. In addition, the M170, generating from the right lateral occipital cortex and the right fusiform gyrus, have similar features in individuals with congenital prosopagnosia (Rivolta et al. 2012b; Towler et al. 2012). Taken together, these studies suggest a prototypical functioning of the posterior face regions in congenital prosopagnosia[3] and points towards other brain regions as possible neural substrates of congenital prosopagnosia (Avidan and Behrmann 2009).

---

[3] We cannot, however, exclude that MEG activity within the right lateral occipital lobe and the right fusiform gyrus will show abnormal features when considering components occurring later than the M170, such as the M400. Future research will clarify this issue.

An additional region critically involved in face processing is the anterior temporal lobe (AT) (Kriegeskorte et al. 2007; Rajimehr et al. 2009; Williams et al. 2006). Previous investigations of the anterior temporal regions in congenital prosopagnosia revealed some significant differences compared to "normal" individuals. For instance, compared to matched healthy participants, people with congenital prosopagnosia exhibit a volume reduction in the AT, and this reduction is correlated with behavioral performance: as AT volume decreases, so does behavioral performance (Behrmann et al. 2007; Garrido et al. 2009). As discussed, recent research has demonstrated reduced structural connectivity (white fiber connections) between posterior and anterior brain regions in congenital prosopagnosia (Thomas et al. 2009). This reduced structural connectivity may be causing the poor connection between posterior and anterior face regions. Finally, preliminary evidence suggests that the pattern of neural activity within the anterior temporal cortex is less face-specific in congenital prosopagnosics compared to "normal" individuals (Avidan 2012; Rivolta et al. 2011).

Together, the findings summarized above strongly support a neuro-anatomical model of congenital prosopagnosia, with a particular emphasis on the abnormal structure, connectivity and functioning of the AT (see Fig. 3.8). Is this neuro-anatomical model of congenital prosopagnosia in agreement with the behavioural findings? Despite impaired overt recognition skills in congenital prosopagnosia (e.g., difficulty in recognizing people by their faces), behavioural findings suggest covert face recognition when, for instance, completing the Forced choice familiarity task; however, no covert processing was evident in the Priming task.

The neuro-anatomical model of congenital prosopagnosia can account for all these behavioural findings: the above-chance covert discrimination between familiar and unfamiliar faces on the Forced choice familiarity task might be facilitated by normal functioning of posterior (core) face areas, since the distinction between famous versus unfamiliar faces is mediated, at least in early processing stages, by neural activity within the occipital lobes. In contrast, access to specific semantic/biographical information, including names, relies on anterior temporal regions (Gorno-Tempini et al. 1998; Palermo and Rhodes 2007; Seidenberg et al. 2002). The anatomical/functional abnormality of the AT and its disconnection with posterior face regions suggests that tasks requiring the access to semantic/biographical representations (represented within the AT) should show abnormal performance in congenital prosopagnosia. The ability to access semantic/biographical information about familiar faces is crucial for tasks requiring (overt) familiar face identification and on the Priming task. In overt face identification tasks such as the MACCS Famous Face Task 2008 (MFFT-08, see Chap. 3), a seen face must activate specific semantic/biographical representations in order to be identified. In the Priming task, access to semantic/biographical information about the "prime" face must take place in order for the "target" name to be categorized as an actor or politician. In agreement with the model, congenital prosopagnosics exhibit impairment on the MFFT-08 and demonstrate no evidence of covert priming (Rivolta et al. 2012a).

Taken together, the evidence, as summarized above, strongly support the proposed neuro-anatomical model, which posits that congenital prosopagnosia is the result of an abnormal structure, connectivity and functioning of the AT, not (or to a lesser extent) the posterior face regions.

References

Avidan, G. (2012). *Selective dissociation between core and extended regions of the face processing network in congenital prosopagnosia* Paper presented at the Annual meeting of the Society for Neuroscience, New Orleans.

Avidan, G., & Behrmann, M. (2008). Implicit familiarity processing in congenital prosopagnosia. *Journal of Neuropsychology, 2*, 141–164.

Avidan, G., & Behrmann, M. (2009). Functional MRI reveals compromised neural integrity of the face processing network in congenital prosopagnosia. *Current Biology, 19*, 1–5.

Avidan, G., Hasson, U., Malach, R. l, & Behrmann, M. (2005). Detailed Exploration of Face-related Processing in Congenital Prosopagnosia: 2. Functional Neuroimaging Findings. *Journal of Cognitive Neuroscience, 17*, 1150–1167.

Avidan, G., Tanzer, M., & Behrmann, M. (2011). Impaired holistic processing in congenital prosopagnosia. *Neuropsychologia, 49*, 2541–2552.

Barton, J. J., Cherkasova, Mariya V., & Hefter, Rebecca. (2004). The covert priming effect of faces in prosopagnosia. *Neurology, 63*, 2062–2068.

Bate, S., & Cook, S. J. (2012). Covert recognition relies on affective valence in developmental prosopagnosia: Evidence from the skin conductance response. *Neuropsychology*. doi:10.1037/a0029443.

Bauer, R. M. (1984). Autonomic recognition of names and faces in prosopagnosia: a neuropsychological application of the guilty knowledge test. *Neuropsychologia, 22*, 457–469.

Behrmann, M., Avidan, G., Gao, F., & Black, S. (2007). Structural imaging reveals anatomical alterations in inferotemporal cortex in congenital prosopagnosia. *Cerebral Cortex, 17*, 2354–2363.

Eimer, M., Gosling, A., & Duchaine, B. (2012). Electrophysiological markers of covert face recognition in developmental prosopagnosia. *Brain. 135*(2):542–554.

Garrido, L., Furl, N., Draganski, B., Weiskopf, N., Stevens, J., Chern-Yee Tan, Geoffrey, Driver, J., Dolan, R. J., & Duchaine, B. C. (2009). Voxel-based morphometry reveals reduced grey matter volume in the temporal cortex of developmental prosopagnosics. *Brain, 132*, 3443–3455.

Gauthier, I., Tarr, M. J., Moylan, J., Skudlarski, P., Gore, J. C., & Anderson, A. W. (2000). The fusiform "face area" is part of a network that processes faces at the individual level. *Journal of Cognitive Neuroscience, 12*, 495–504.

Gorno-Tempini, M. L., Price, C. J., Josephs, O., Vandenberghe, R., Cappa, S. F., Kapur, N., et al. (1998). The neural systems sustaining faces and proper-name processing. *Brain, 121*, 2087–2097.

Gosling, A., & Eimer, M. (2011). An event-related brain potential study of explicit face recognition. *Neuropsycholgia, 49*, 2736–2745.

Haxby, J. V., Hoffman, E. A., & Gobbini, M. I. (2000). The distributed human neural system for face perception. *Trends in Cognitive Sciences, 4*, 223–233.

Jones, R. D., & Tranel, D. (2001). Severe developmental prosopagnosia in a child with superior intellect. *Journal of Clinical and Experimental Neuropsychology, 23*, 265–273.

Kanwisher, N., McDermott, J., & Chun, M. M. (1997). The fusiform face area: A module in human extrastriate cortex specialized for face perception. *Journal of Neuroscience, 17*, 4302–4311.

Kempler, D. (2005). *Neurocognitive Disorders in Aging* (1st ed.). Thousand Oaks: Sage Publications, Inc.

Kriegeskorte, N., Formisano, E., Sorger, B., & Goebel, R. (2007). Individual faces elicit distinct response patterns in human anterior temporal cortex. *Proceedings of the National Academy of Science USA, 104*, 20600–20605.

Marshall, J. C., & Halligan, P. W. (1988). Blinsight and insight in visuo-spatial neglect. *Nature, 336*, 766–767.

Palermo, R., & Rhodes, G. (2007). Are you always on my mind? A review of how face perception and attention interact. *Neuropsychologia, 45*, 75–92.

Palermo, R., Willis, M. L., Rivolta, D., McKone, E., Wilson, C. E., & Calder, A. J. (2011). Impaired holistic coding of facial expression and facial identity in congenital prosopagnosia. *Neuropsychologia, 49*, 1226–1235.

Rajimehr, R., Young, J. C., & Tootell, R. B. H. (2009). An anterior temporal face patch in human cortex, predicted by macaque maps. *Proceedings of the National Academy of Science USA, 106*, 1995–2000.

Renault, B., Signoret, J.-L., DeBruille, B., Breton, F., & Bolgert, F. (1989). Brain potentials reveal covert facial recognition in prosopagnosia. *Neuropsychologia, 27*, 905–912.

Rivolta, D., Palermo, R., & Schmalzl, L. (2013). What is Overt and what is Covert in Congenital Prosopagnosia? *Neuropsychology Review, 23*(2), 111–116. doi:10.1007/s11065-11012-19223-11060.

Rivolta, D., Palermo, R., Schmalzl, L., & Coltheart, M. (2012a). Covert face recognition in congenital prosopagnosia: A group study. *Cortex, 48*, 344–352.

Rivolta, D., Palermo, R., Schmalzl, L., & Williams, M. A. (2012b). Investigating the features of the M170 in congenital prosopagnosia. *Frontiers in Human Neuroscience, 6*, 45.

Rivolta, D., Schmalzl, L., Coltheart, M., & Palermo, R. (2010). Semantic information can facilitate covert face recognition in congenital prosopagnosia. *Journal of Clinical and Experimental Neuropsychology, 32*, 1002–1016.

Rivolta, D., Schmalzl, L., Palermo, R., & Williams, M. A. (2011). *Multi-voxel pattern analysis of fMRI data reveals abnormal anterior temporal lobe activity in congenital prosopagnosia*. Paper presented at the European Conference on Visual Perception (ECVP) Toulouse.

Schmalzl, L., Palermo, R., Green, M., Brunsdon, R., & Coltheart, M. (2008). Training of familiar face recognition and visual scan paths for faces in a child with congenital prosopagnosia. *Cognitive Neuropsychology, 25*, 704–729.

Schweinberger, S. R., & Burton, A. (2003). Covert recognition and the neural system for face processing. *Cortex, 39*, 9–30.

Seidenberg, M., Griffith, R., Sabsevitz, D., Moran, M., Haltiner, A., Bell, B., et al. (2002). Recognition and identification of famous faces in patients with unilateral temporal lobe epilepsy. *Neuropsychologia, 10*, 446–456.

Thomas, Cibu, Avidan, Galia, Humphreys, Kate, Gao, F., & Behrmann, Marlene. (2009). Reduced structural connectivity in ventral visual cortex in congenital prosopagnosia. *Nature Neuroscience, 12*, 29–31.

Towler, J., Gosling, A., Duchaine, B., & Eimer, M. (2012). The face-sensitive N170 component in developmental prosopagnosia. *Neuropsycholgia*, http://dx.doi.org/10.1016/j.neuropsychologia.2012.10.017.

Williams, M., Savage, G., & Halmagyi, M. (2006). Abnormal configural face perception in a patient with right anterior temporal lobe atrophy. *Neurocase, 12*, 286–291.

Chapter 5
Stories from People Who Share Their Lives with Congenital Prosopagnosia

5.1 Introduction

Up to now we have learned about cognitive science and its methods, about the cognitive and neural features of typical face processing, about congenital prosopagnosia and the mechanisms behind covert face recognition. What we have overlooked, so far, is the point of view of people that live with prosopagnosia every day. What is it like living with congenital prosopagnosia? What are their experiences, their feelings and thoughts about their problems? What are their hopes for the future?

Below are some stories of adults and children with severe problems in the ability to recognize faces who, over the last 5 years, have collaborated with great patience and dedication in our research projects. For obvious privacy reasons their names have been changed; however their biographical and psychological features reflect the truth. The first part of the chapter reports cases of adults with prosopagnosia, whereas the second part describes cases of kids with face recognition difficulties. These last cases were reported and shared by a former colleague at MACCS, Dr Catherine E. Wilson.

5.2 Cases of Adult Congenital Prosopagnosia

Tiffany

Background. Tiffany is a 50 year-old Australian woman who lives in a small town few hours drive from Sydney, in New South Wales. There she is employed as a saleswoman in a big shop which sells clothes and house products. In 2007, after watching a TV program where some researchers from MACCS described congenital prosopagnosia, she decided to contact us in order to ascertain whether she was prosopagnosic or whether her face recognition impairments had a different origin.

This chapter was written in collaboration with Dr Catherine E. WIlson.

D. Rivolta, *Prosopagnosia*, Cognitive Systems Monographs 20,
DOI: 10.1007/978-3-642-40784-0_5, © Springer-Verlag Berlin Heidelberg 2014

During the first interview, Tiffany appeared very motivated in understanding what was going on with her. We administered a neuropsychological battery, which indicated that her intelligence, general cognitive function, general vision, social skills and object recognition all were within the normal range. However her ability to recognize famous faces (MFFT-08) and to remember previously unknown faces (CFMT) were below the norms expected for her age and sex. This, along with the absence of any evident sign of brain lesion led us to formulate a "diagnosis" of congenital prosopagnosia. This was also strengthened by the fact that, in her opinion, also her mother, some cousins and one of her sons present similar difficulties with faces. Below are some extracts of her stories.

Childhood and Adolescence. My earliest memory of a sign of CP was when I was about 7 years old. There was a low foot bridge which my siblings and I crossed to get from our home to school. It had been washed away during heavy rain, so we were forced to walk about 2 km to another bridge. A man driving past stopped and offered us a lift. My siblings accepted, but I refused as I thought he was a stranger. It turned out he was the priest at the church we all attended each week. I had known him for 2 years. I didn't recognise him because he was in normal clothes.

[…]

I attended a small school in a rural town from the age of 5–11 years. There were only about 100 children in the entire school, about 18 in my class. When I was in 5th Grade, my teacher asked me to hand back my classmates' books which he had been marking. I could only match about 6 of the names on the books to the children sitting at the desks. I had shared a class with these same children daily for 5 years! I had no problem with the girl with long, red hair, or the one with bright white hair, but at least 6 of the other girls looked identical to me. It was even worse with the boys, because they all had short hair.

[…]

My family moved to Sydney at the same time I started high school. I babysat for a few families as a part time job. One family I looked after at night, while both parents were working shift work. I looked after their two children four nights per fortnight. One day I was shopping with my sister at the local shopping centre. A family was walking towards us, so I moved to the side to let them pass. My sister said to me, 'What's wrong? Aren't you going to say hello to them?' She had to tell me it was the family I worked for. I didn't recognise them outside their home. I had been working for the family for over a year!

[…]

One day my brother was playing at home with his friend Richard, who was our neighbor. A lady came to the door, so I assumed it was Richard's mum asking for him to come home. It turns out it was a lady from further up the street, wanting me to babysit that weekend. I had known both ladies for about 6 years. I could never tell them apart.

[…]

I always had just one or two friends at school. They tended to be unpopular with the rest of the class. I think I subconsciously hung around with kids who looked very different to the others. In high school my friends for the first 3 years were the tallest girl in the class and the fattest girl in the class. I don't believe I made these choices deliberately. In my last 3 years of high school, my friends were the only Chinese girl in the school and an English girl with curly hair.

[…]

Just after I finished high school, I was walking through the shopping centre when I was approached by three people wearing afro wigs. I knew they were three people I had attended school with for 6 years, but I had no idea who they were. They were joking about saying, 'I bet you don't know who we are,' because of the wigs. I was very embarrassed and lied, 'Of course I know you,' then ran away from them.

Adulthood. I work in a department store. I have been there over 6 years. Fortunately for me, we are required to wear name badges. This should make life much easier for me. Unfortunately, some employees forget to wear their badges, or have them obscured under their jacket or cardigan. They are often carrying stock in their arms when they stop to ask something. This also hides their name. Occasionally a manager will check that everybody has a name tag. If someone has forgotten theirs, they will borrow one from a colleague who has just finished their shift, so they have the wrong name on. You can imagine how confusing this is for me! Fortunately for me, I can recognize our Store Manager, because there are only three men employed during my shifts, and the manager is at least 8 cm taller than the other two!

[…]

It is a small store, and during my shift I only have to interact with about twelve to fifteen co-workers. After 6 years I can recognize 5 people if I meet them out of uniform. There are 4 or 5 young girls who all look identical to me, and they keep changing their hair! I have trouble telling who is who when they are together at work, but if I meet any of them elsewhere, I don't even know I've met them before. There are two older ladies who have similar hairstyles. I can only tell them apart if they are standing next to each other. There are four middle-aged ladies who all have a similar body shape. I'm not sure which is which even when they are standing in a group. Where possible, I use their name badges to help me, but, as I've explained, this is often impossible.

[…]

One morning per week, I am the 'Door Greeter'. I let the employees in the staff entrance before the store opens, then I stand at the main door, greet customers, answer enquiries and do security checks. When the staff comes through the staff entrance, they usually don't have their name badges on, because most people keep them in their locker. Sometimes a manager will ask me questions such as, 'Has Mary come in yet?' or 'Who's come in for Home Entertainment, Jane or Sue?' Somehow I get by with giving a vague answer, such as, 'I'm not sure, she may have come in when I was talking to Anne.' Every time this happens, I feel so embarrassed. It upsets me, because I want to be seen as a competent employee, not an idiot.

After the store opens, while I am dealing with customers at the main door, other members of staff come into start later shifts. One day the Store Manager advised me that the Regional Manager and Area Manager would be paying us a visit. He asked me to advise any staff who were coming on shift to make sure they were wearing their name badges and that they kept their areas tidy. A young lady came in wearing our uniform, but without her badge. I told her, 'You'd better put on your badge. The bosses are coming for a visit.' It was the Area Manager! I didn't even realize that it was someone I'd never worked with before! She just said, 'Oh, I'd better do that.' I hope she thought I was just having a joke with her.

Another day, I served a lady who was buying a pair of shoes. There was a problem with the price of the shoes, so I asked the supervisor to call an assistant from the shoe department. Someone came, I gave her the details and she went to find the correct price. After about 5 min, the customer was still waiting for an answer. I advised the supervisor, who asked me, 'Who came down to check on the shoes, Kim or Lisa?' I decided to take a chance and answered, 'Lisa.' Just then Kim came back with the shoes!

On another occasion I served a man at the checkouts. Just after he left, I realized he had left a bag of items behind. I told the supervisor, who grabbed the bag and started running to the door. There were two men leaving. She asked me which man I had just served. I didn't know.

Again, as a lady came into the store, I started to tell her about a special sale in the ladies wear department. She gave me a strange look and kept walking. I later realised she was the manager from that department! She was not in uniform because it was her day off.

[…]

I volunteer with an auxiliary group which supports the local hospital. I have been with the group for 9 years. There are seven regular ladies who volunteer on the same day of the week as I do, including the coordinator. We all wear a pink uniform. One evening I was invited to a fund raising event with the coordinator. I had to meet her at a restaurant. She arrived before me. I knew she would not be wearing her uniform and worried that I might not recognize her. I entered the restaurant hoping that she would see me first and call out to me, so I would know who she was. I saw a lady about the correct age and height, but unfortunately she was talking to the man who had arranged the event and she didn't notice me. I couldn't recognise her without her pink uniform. I was too embarrassed to approach her in case it was the wrong lady. I walked back outside and watched through the window until she was alone, then I walked back inside. The lady noticed me and called out, 'Hello', so I knew she was the correct person. I have worked with her for at least two hours per week for the past 7 years.

A lady came up to me at the store where I work and started chatting with me as if we were friends. I vaguely recognized her, but I couldn't work out where I knew her from. After she talked a bit longer, I realized she was one of the Pink Ladies who I have morning teas with every Tuesday.

[…]

My younger son played soccer from about 6 years of age to 11 years of age. He was in the same team for all those years with mostly the same group of boys. They were all in his class at school. I took him to soccer practice for 1 h every Tuesday afternoon and to soccer games every Saturday morning. While he was on the field, I chatted to the parents of the other boys. I helped them with fundraising and volunteered at the canteen. I also saw them at school events.

After all those years I can only recognize one of the mothers, because she has never changed her hairstyle. I can also recognize two of the fathers. I still can't tell who is who between the team manager and the coach. I was never able to recognise any of the boys, except the tallest one, when he was with the rest of the team. If he was with his older brother, I couldn't tell them apart. If I saw any of the boys at school, I didn't even realize they were members of my son's team.

[…]

We paid a small fee each week, which was collected at the soccer game. This was usually collected by the coach's wife, who collected the money and ticked off each name on a list, to show we had paid. One day she asked me to collect the fees for her. This meant I had to identify the parents of my son's team mates among the crowd of people watching the game, so that I collected money from each family, then tick off the correct name! These are people I had known for over 5 years, and I realised I could only recognise and name 3 of them. I approached the couple of people I knew, then made some excuse for someone else to take over. I was so upset, because I like to help out and I felt I was letting the team down.

[…]

At the end of each soccer season we had a party with the team and parents. The party was held at a different home every year, and each family brought along food to share. One year, a few days after the party, I realised that I had left behind the crystal platter on which I had taken food. I didn't socialise with these families except for soccer events, so I knew I would not be going back to that family's home. When I next met with the soccer team, I couldn't recognize which mother had hosted the party, so I couldn't ask for my platter to be returned.

[…]

I attend an evening class once a week at the local TAFE (Technical College). One evening, when I had been in the same class, with the same teacher and students, for about 6 months, I arrived a few minutes late. I was about to take my seat when I noticed that there was a different teacher. I hadn't noticed that the entire class was strangers. My class had moved to a different room that night.

[…]

On most afternoons I drive to the local stables to pick up my daughter, who keeps her horses there. While I am waiting, I see many women walk past my car to feed their horses. There are at least 3 middle-aged ladies who all look the same to me. I have met these ladies many times but I don't know which one is which. If they come into the store where I work, I don't even know I've met them before. If a lady comes in and starts chatting about my daughter and horses, I assume it is one of these ladies, but I'm usually too unsure to join in the conversation.

There are also about 4 younger ladies who visit the stables regularly. I can't tell them apart.

Until recently, there was only one man who kept horses at the stables. I felt quite comfortable chatting to him at the stables, because I knew who he was, as he was the only male. One afternoon I saw a man walk past my car to the stables. I assumed it was Jim and started chatting to him as usual. He seemed a little distant, and continued on to the stables. Five minutes later Jim drove up and parked next to my car! It turned out that the man I had been speaking to earlier was Anthony, who had just moved his horses to the stables that week. Both Jim and Anthony are middle-aged and balding. I am now anxious every time I go to pick my daughter up at the stables. If I see one of the men, I am not sure whether to start a conversation or not. If it is Jim and I don't talk, he might think I'm being rude, but if it's Anthony, he will be confused if I start a conversation with him. I find these situations very upsetting.

[…]

As I was finishing work one day, one of my workmates told me that there were some sales in the toy department that I might be interested in. She pointed out a specific item which I had wanted to buy for my son. It was reduced by 60 %. I thanked her, and bought the item. That evening I gave the item to my son and he was very excited to finally have it. When I was on Facebook, I thought it would be nice to message my workmate with a photo of my son holding the toy, to show how grateful we were that she had remembered that he wanted it. Unfortunately, I couldn't work out whether it was Robyn or Sandra that had helped me. I was about 80 % sure it was Robyn, but I didn't want to embarrass myself by thanking the wrong person. As usual, I just did nothing. I feel really bad, because whether it was Robyn or Sandra, either of them would have been happy to receive the photo, but I didn't want to send it to the wrong person.

[…]

There are a few older ladies who come to the store regularly. Usually it doesn't matter that I can't tell them apart, because I just greet them with a friendly smile and help them find something if necessary. One lady, however, recently told me that she had lost her husband suddenly. Now where an older lady comes in, I feel I can't greet her with my usual friendly smile in case it is the lady who is in mourning. I would love to be able to greet that lady compassionately and ask how she is getting on. I'd like to show her that I care about what she is going through. I certainly don't want to seem insensitive by greeting her with a jovial smile. So, since I don't know which lady she is, I am greeting all the older ladies a little more formally, just in case. I hope the other ladies don't think I'm being unfriendly.

[…]

By nature I am a friendly, helpful person. I want to be the person who greets regular customers by name. I rarely greet anyone by name, in case it's not the person I thought it was.

I want to recognise the customer who was looking for green wool the previous week, and let her know that it is now in stock.

If I serve a customer who is buying items for a vacation, I want to recognise them in a month's time and ask them how they enjoyed their trip.

If I am walking down the street and a lady approaches me and starts talking to me, I want to know instantly that she is the mother of my son's friend. I want to recognise her, greet her by name, ask her how her son is doing in his new school.

I want to ask her how her husband's knee surgery went, and tell her that I like her new hairstyle. My reality is that if I am walking down the street and a lady approaches me, I feel overwhelming stress. Is this someone I know from work, from my son's school, from my daughter's stables? Does she know me well, or has she just seen me in the store where I work? I let her take the lead in the conversation, hoping that what she says and asks will give me some clue as to who she is. The whole time I'm listening to her speak, I'm worrying that I will appear unfriendly or abrupt, because I am not joining in the conversation in a natural manner.

Jen

Background. Jen is a 44 years old woman that lives near Sydney, in Australia. She is married with 2 kids, she has a Ph.D. in pharmacology and she now works in the private sector. The neuropsychological assessment showed normal cognitive general functioning and intelligence (she is a doctor!), normal memory for objects and normal vision. Her only complaint was about her poor face recognition skills. She was so interested in finding out more about prosopagnosia that she agreed to get all her family (mum, husband and the two kids) tested for prosopagnosia. Results, in agreement with scientific literature, indicated a familiar component: her 6 year-old son, Paul, is also prosopagnosic (we will discuss in detail Paul's case in the next section).

Below is what she told me about her everyday-life episodes indicating congenital prosopagnosia.

"About a month after I met my husband-to-be Max, the two of us went to one of his friend's wedding. I am an extrovert and love to chat so it didn't take long before I found myself having a conversation with one of the male guests and his girlfriend. We stayed on friendly terms with the couple all night. On our way home, Max looked a little confused and asked whether the male guest we were chatting to was the same person I had gone out with for about 2 weeks just before Max and I had become a couple. At first I disagreed, after all, I'd recognise someone I'd gone out with for 2 weeks. However, after about half an hour of discussion, it dawned on both of us that Max was absolutely right.

I was so embarrassed that I never really thought about this event again until I heard about prosopagnosia during a TV show which was presenting Macquarie University's research on the topic. The University was asking for volunteers to participate if they felt that they might have the condition. I put up my hand and it soon became clear that I did indeed have prosopagnosia. It also began to clarify why all through my own university days I would meet people on trains who were certain that they knew me but I wouldn't have a clue who they were. It happened so often that I eventually decided that I had a doppelganger at Sydney University and that people were mixing me up with someone who must have looked a lot like me. In hindsight, they were probably right, they did know me it was just that I didn't recognise them!

Max and I began to suspect that my eldest child, Paul, also had prosopagnosia when he was still at pre-school. There were little hints, like he seemed to get his teachers confused and he never seemed to know the names of any of his friends. We put this down to being young and as he was our first-born we didn't know any better. We became convinced though when a good friend of his, Caitlyn, got her hair cut. Caitlyn and Paul had known each other since they were both babies as our families belonged to the same parent/mothers group. They loved hanging out with each other, dancing at festivals and playing both at pre-school and on play dates. At one point, Paul was so enthusiastic about Caitlyn that he gave her such an almighty hug he managed to knock them both over a park bench, after which Caitlyn was a little less enthusiastic about Paul.

However, at about 4 years of age they seemed to quite suddenly drift apart. Paul didn't seem to talk about Caitlyn and at the park they both played with other friends whilst barely talking to each other. We kept asking Paul what he thought of Caitlyn's new haircut but he didn't really respond except to say he liked girls with long hair. We thought it was very odd that our wonderfully friendly and egalitarian young son seemed to have gone off Caitlyn purely because she no longer had long hair. It was about this time that I learned about my own prosopagnosia and had a hunch that this might be behind Paul's unfriendliness. It took a while, but we finally worked out that Paul had decided that Caitlyn with the short hair was "not the same Caitlyn" that he once knew. Unfortunately, by this time we had worked this out their closeness had been lost although I am pleased to report that they seem to be becoming good friends again, now that they are eight. Caitlyn has had lots of different haircuts over the last couple of years but Paul seems to be recognising her regardless of her hairstyle. It has been fascinating to watch how different Paul and his younger brother Greg are. Paul can never tell us who he is in a group with or which teacher said what. Greg on the other hand seems to know the names of almost every child in the school and rarely gets his teachers mixed up. Prosopagnosia testing is consistent with what we guessed—Paul has prosopagnosia whilst his younger brother doesn't appear to. Despite this, Paul is by far the more social child and is generally liked by all who meet him. His natural friendliness makes up for his deficiency in recognising people. Perhaps it helps him forgive more easily?

5.3 Cases of Children with Face Recognition Difficulties

Assuming that congenital prosopagnosia affects 2–3 % circa of the general population, it is curious to note how only five cases of kids with this condition have been reported in scientific literature before 2010 (Ariel and Sadeh 1996; Brunsdon et al. 2006; Jones and Tranel 2001; McConachie 1976; Schmalzl et al. 2008). On closer inspection, none of these reports could confidently claim that the children in question had problems with face recognition that were not due either to a general visual processing deficit, or to a more global developmental disorder. One explanation for the lack of childhood cases reported is that face recognition problems in children go unnoticed. If a child writes their letters backwards or develops a speech impediment, the problem would, one hopes, be picked up and addressed by their parents and teachers. Face recognition skills, however, are not taught or tested, and neither can a child easily compare their skill level to that of their peers. Moreover, there are numerous ways to compensate: Beth might correctly identify that boy as Johnny because Johnny always sits in that seat, or has red hair, or always makes a lot of noise. If Beth relies on non-facial cues to identify her peers, why should she realise that she is different to anyone else?

An alternative explanation for the lack of childhood DP cases is that face recognition deficits do not develop in isolation in childhood, but develop only in the context of a general visual processing impairment, or when the child has a more global developmental disorder of which face recognition problems are one symptom. The developmental disorder that is of particular relevance here is autism spectrum disorder (ASD), which is diagnosed when impairments in social interaction and communication co-occur with repetitive behaviours and restricted interests. Concerns regarding the presence of ASD symptoms in children with congenital prosopagnosia are pertinent because face recognition impairment is a common symptom of children with ASD. In this section we describe five cases of potential childhood DP that were brought to our Face Recognition Research Group by concerned parents. We used a number of tasks, across two different sessions, to test the children's immediate face recognition (recognising unfamiliar faces a few seconds after seeing them) and memory for faces (recognising unfamiliar faces over a period or 30 min), and compared scores to a group of age matched controls (see Fig. 5.1 for an example of stimuli adopted).

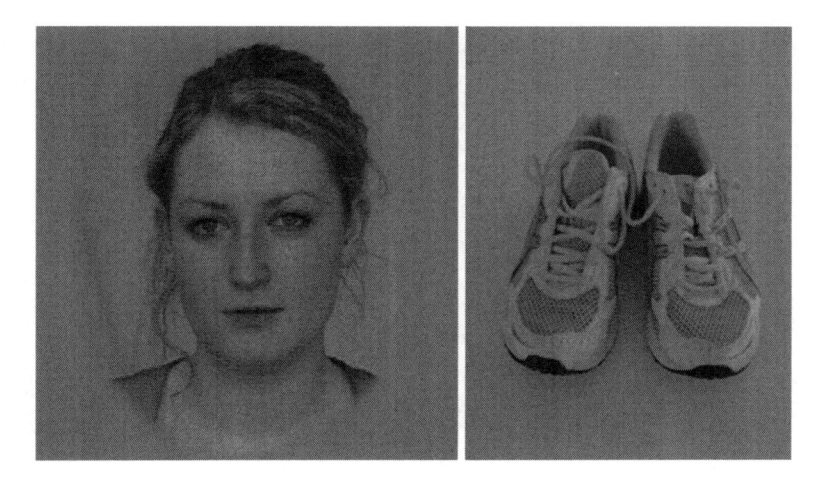

Fig. 5.1 Examples of stimuli adopted for the analysis of cognitive abilities in people with suspect prosopagnosia. Memory is assessed with face (*left*) and shoes (*right*) in order to understand whether the eventual problem is face-specific or not

Verbal and non-verbal IQ was also tested, and found to be well within the normal range for each child. In addition, we set out to eliminate the possibility that they were on the autistic spectrum, or that they had more general visual processing deficits, in an attempt to identify a genuine case of childhood congenital prosopagnosia.

Kate

Kate was 7½ years old when her mother brought her to the University to meet me. She was shy at first, but after some reassurance she allowed her mother to leave the room while she completed her tasks. Kate was diagnosed with Poland's syndrome as an infant—a congenital condition described as an underdevelopment or absence of the chest muscle on one side of the body, and she also appeared to have slightly dysmorphic features. As a young child, a number of unusual behaviours were noted: toilet training was problematic and delayed, temper tantrums were common, she was terrified of other children and suffered from severe separation anxiety. Kate's parents decided to home-school her to spare her the anxiety she suffered when in situations with several children. An attempt to take her to 'Brownies' induced such tension, stress and anger that her mother gave up after two sessions. The possibility of autism was raised when she was three, but diagnostic screening concluded that she was not on the spectrum.

Kate was 4 years old when she was first referred for a general neuropsychological assessment because of apparent difficulties in recognizing familiar faces. Kate tended to rely on clothing to identify others, and once failed to recognise her own father when he was wearing the same football strip as his team-mates. Kate's parents both had good face recognition abilities, although prosopagnosia had been confirmed in other family members (cousin, uncle, 2 grandparents and great-grandmother).

Kate's performance on the tasks confirmed that she struggled with immediate face recognition, although when she had more time to remember them she performed only slightly below what was expected for children of her age. A comprehensive interview to assess autistic traits confirmed that Kate did in fact meet criteria for an autism spectrum disorder.

The conclusion? Face recognition problems were part of more general developmental difficulties, best described as autistic spectrum disorder.

Samuel

I visited Samuel in his home the day after his 6th birthday and found him engrossed in a Lego pirate set he had been given as a present. With some encouragement he completed all the tasks I set him, so long as he was allowed to fit a few more pieces of his Lego set in between tests. When I arrived the following day to complete the session, Samuel looked perplexed and whispered something to his mother—he asked her if I was the same lady that had visited the day before. On being assured that I was the same person, he wanted to show me the vast and intricate pirate model that he had worked on tirelessly and completed after I left the previous day.

Samuel was born 6 weeks premature, was severely delayed in developing language skills, and did not start to talk until he was 4 years old. Concerns about Samuel's development led his parents to have him assessed for autistic spectrum disorder when he was 3 years old, the conclusion being that he suffered from a pragmatic language disorder.

Samuel's parents told me that he often failed to identify family members. Once, he quite earnestly asked his father in a supermarket "are you my daddy?" which understandably caused his parents some concern: if he couldn't recognise his father, how was he coping in other situations? Apparently, not that well. After 3 months of starting a new school the only child he could confidently name was the only child of Asian origin. Samuel was drawn towards peers that were visually distinctive, for example two girls with long red hair (who, incidentally, he often confused).

On the tasks we set him, Samuel performed very poorly when unfamiliar faces had to be recognised immediately[1] (lowest 5th percentile). Conversely, his memory for unfamiliar faces was within the normal range. However, an interview for autistic traits with his parents suggested that he met criteria for autistic spectrum disorder. His poor eye contact, lack of interest in other children, lack of imaginative play and restricted interests (e.g. in model toys) were among the traits that placed him on the autistic spectrum.

The conclusion? Samuel was on the autistic spectrum, therefore his face recognition problems were considered to be part of more a general developmental disorder.

Paul

Paul was just short of 6 years old when I visited him at his home. He was immediately chatty and uninhibited, and keen to get started on his tasks (his enthusiasm fuelled by the fact that the lady with the computer had come to visit him and not his little brother). Paul's mother described him as a highly sociable child who was friendly to everyone. In fact, he seemed equally happy to play with children he'd never met before as he was with those that he went to school with. Despite this, his social skills were adept for a 6 year old, with the exception that he tended to avoid direct eye-contact. Paul also suffered from childhood arthritis.

Paul's mother became aware of the difficulties he was having after a number of incidents at school. For example, Paul didn't notice that his regular teacher had been replaced by a substitute for several days. One of his friends cut her hair short and Paul had no idea who she was, and still did not believe it even after being

---

[1] In psychology we call *short term memory* the memory we use to retain information for short periods of time, such as when we read and soon after digit a phone number, and *long term memory* when the one we use to store and retrieve events that occurred last night or even many years ago.

told. He often talked about things other children had said or done at school, but would not refer to them by name. When asked who he was talking about he would say 'the boy with the red hat' or 'the girl with the curly hair'. Paul's mother (Jen, see previous section), who suffers from face recognition difficulties herself, recognised this habit as something she often did herself.

Our tests confirmed that Paul had severe difficulties with immediate recognition of faces, as well as being very poor at remembering new faces over short time periods (lowest 5th percentiles). However, Paul also struggled with recognising other objects, performing in the lowest 10th percentile for his age group despite his normal eye-sight and his average intelligence. As expected, traits of autism were well below threshold in Paul's case.

The conclusion? Face recognition problems were accompanied by more general difficulties with visual processing.

Nick

Nick was 7½ years old when I went to visit him in his home. He was apprehensive, but polite, and his mother informed me that he was usually shy around adults. Recently, Nick avoided saying 'hello' to his aunt and uncle when he saw them for the first time in a year, but it then became clear he did not recognise them and thought they were strangers. His mother thought he had difficulties with face recognition from an early age, which she noticed because she also suffered from face recognition problems. Nick's face recognition had not been tested before, but visual spatial skills were tested by an optometrist who found that Nick was in the lowest 10th percentile for his age at discerning the proper orientation of letters and numbers, (assessed using Gardner's Reversal Frequency). Nick also had motor dyspraxia[2] during the first years of his life.

Results of tests showed that Nick's immediate face recognition and his memory for faces were poor (lowest 5th percentile). We tested Nick's recognition skills for objects, this was also found to be impaired. However, in Nick's case, there was no evidence of autistic traits.

The conclusion? Face recognition problems were part of a more general difficulty with visual processing.

Andy

Andy was nearly 8½ years old when he came to our University to be tested. He was boisterous, confident, and couldn't wait to get started on the computer 'games'. His mother commented that he was a sociable child and that he was different,

---

[2] Deficit in movement organization (e.g., clumsiness) which follows a central problem and not effector muscles.

unconventional, 'quirky'. When he was 4 years old his kindergarten teacher mentioned that he often failed to recognise the children and adults that he saw regularly. The possibility of prosopagnosia was raised at the time, although no formal testing was conducted. His mother reported that each year when he starts a new class it takes him a long time to be able to identify the other children, and he tends to single out individuals that are distinctive (e.g. a girl with very long hair) and play more one-to-one than in groups. He was also happy to spend time alone, getting absorbed in one activity and not paying attention to other people. Andy reached all his developmental milestones on time, although he has poor eye contact particularly with people he does not know. Andy's parents had no face recognition difficulties themselves.

Our tests confirmed that Andy's immediate recognition of faces was poor, and that his memory for faces was severely impaired (lowest 1st percentile). By contrast, Andy performed slightly above average for his age on the tests of object recognition. To establish his impaired memory was for faces specifically we administered tests of general memory function, which confirmed that he could complete working memory tasks at an age-appropriate level, as well as remembering visual and auditory information normally. Anecdotes and descriptions of Andy's character rightly raised concerns that he might have an autistic spectrum disorder, however a comprehensive interview with his mother concluded that Andy's difficulties were not in keeping with a diagnosis of autistic spectrum disorder.

The conclusion? Difficulties with facial identity recognition were present without general visual processing problems, and without an autistic spectrum disorder. Andy's difficulties are best explained as a childhood case of "pure" congenital prosopagnosia.

5.3.1 Summary of Cases

All 5 children exhibited face recognition impairments on a range of tasks across 2 testing sessions. Two of the children (Kate and Samuel) were found to exhibit behavioural symptoms indicative of an autistic spectrum disorder. In the remaining three children we found little indication of social abnormalities and the children were clearly not on the autistic spectrum. Two children (Nick and Paul) also performed poorly on the test of object recognition, suggesting their DP may be a component of more general visual recognition difficulties. One child (Andy) exhibited severe problems on the tests of recognition memory for faces, despite normal memory performance for other stimuli and above average object recognition ability.

5.3.2 Face Recognition Impairments and Social Problems

Two children, Kate and Samuel, were found to be on the autistic spectrum. The relationship between characteristics of autistic spectrum disorder and face recognition problems is currently unclear. On the one hand, it is widely argued that difficulties with face recognition in autistic spectrum disorder are secondary to social and perceptual impairments (Elgar and Campbell 2001). Autistic spectrum disorder individuals often lack interest in social stimuli, particularly faces (Klin et al. 2002; Wilson et al. 2012), and it has been suggested that experience with face stimuli during early childhood can affect the development of expert mechanisms for processing facial stimuli (Nelson 2001). Therefore, one possibility is that children with reduced interest in social stimuli may develop atypical or immature face processing skills as a result of reduced experience with faces (Wilson et al. 2011).

However, the two children in question here, Kate and Samuel, had been assessed for autistic spectrum disorder at 3 years of age, but neither was given a diagnosis. Of course, this might reflect a lack of sensitivity of the diagnostic tools used in the earlier assessment, but may also reflect a genuine increase in the severity of autistic symptoms with age (Honey et al. 2008; Moore and Goodson 2003). This raises an alternative possibility: that the face recognition difficulties in Kate and Samuel might have exacerbated and increased the social and communication characteristics of the autistic profile. In congenital prosopagnosia, perhaps, diminished feelings of familiarity, and the resulting curtailment of emotional experience, would have widespread effects on social cognition and functioning, potentially leading to social deficits akin to those experienced in autistic spectrum disorder (Kracke 1994). A more extreme proposition is that deficits processing face information in early childhood are instrumental in causing the social impairments that are central to autistic spectrum disorder (Schultz 2005). Thus, Kate and Samuel's prosopagnosia might have played a causal role in the social impairments of their autistic spectrum disorder (although it is difficult to imagine how face recognition impairments might have contributed to the repetitive behaviours and restricted interests that were also noted in their autistic spectrum disorder symptom profiles).

5.3.3 Face Recognition Impairments and General Visual Processing Difficulties

In two cases, Nick and Paul, performance on a control task using shoes instead of faces was also impaired, suggesting that their face recognition problems were in the context of more general recognition deficits. As noted above, Nick had delayed development of visual spatial skills. However, there was no evidence for similar difficulties affecting Paul, who performed well on a test of visual acuity. Like Nick and Paul, many adults classified as having congenital prosopagnosia also have more general problems of within-category object recognition (Behrmann et al. 2005; Duchaine et al. 2007; Duchaine and Nakayama 2005). Regardless of final classification, these two children could not be considered cases of 'pure prosopagnosia'.

5.3.4 Isolated Face Recognition Problems

In the final case, Andy, some subtle social oddities were reported, however he scored well below threshold on tests of autistic spectrum disorder. He performed slightly above average on the test of object recognition, but he was moderately impaired at immediately recognising faces. In contrast, Andy was severely impaired on two tests of memory for faces, but had no difficulties completing memory tests for other information (visual or auditory). The most appropriate description of Andy's face recognition difficulties seems to be 'prosopamnesia', that is, the ability to perceive identity-related information in novel faces but not to store or consciously retrieve a memory record of facial information (whilst showing normal memory for other material) (Tippett et al. 2000; Williams et al. 2007).

To conclude, closer inspection of these five potential childhood congenital prosopagnosia cases confirmed that face recognition problems in childhood are often accompanied by additional developmental atypicalities. This is unsurprising given that development is an interactive process such that cognitive and perceptual skills may affect, and compensate for one another as they mature. However, face recognition problems *can* be independent from developmental disorders (specifically autistic spectrum disorder) and they *can* develop in the context of normal visual processing. We found one case of "pure" congenital prosopagnosia in childhood, suggesting that even while cognitive systems are still developing, face recognition mechanisms occupy some unique status that has the capacity to be disrupted in isolation.

5.4 Conclusions

The cases described so far make us think on how the face recognition impairment could represent the most significant problem (i.e., "pure" congenital prosopagnosia) or could represent only one of numerous other problems that express themselves within developmental conditions such as ASD.

Stories from Tiffany and Jen suggest that congenital prosopagnosia deeply affects the quality of life of people from childhood until adulthood. It is not rare, unfortunately, to hear stories of social anxiety and avoidance. When I communicate the decision to write this book, Tiffany was extremely happy and relieved; she hopes that this message gets to the highest possible number of people, both prosopagnosics and clinicians, too often unaware of the condition.

From the professional point of view it would be beautiful if 1 day (I hope very soon), we could return the effort and passion these people (and other who participate in our research) put for helping us out, and tell us that we finally got some rehabilitations protocols that can help them ameliorate their condition. The omen is that, not only specialists (clinicians and researchers), but also teachers will be more aware of the condition. To this end, it would be useful an "awareness intervention" at schools, aimed towards the early detection of face skills problems. In fact, even though at the moment there are not "cures" for the condition, just the knowledge that this crazy difficulty in recognize faces has a name, it is quite common, has a genetic component, it does not affect general intelligence or performance at school and that many scientists in the world are studying it, is somehow, by itself, therapeutic.

5.5   What to do in the Event You Believe to have Congenital Prosopagnosia?

You may wonder what to do in the unfortunate event that yourself or someone in your family has problems in face recognition. Who can you contact? Do you need to worry about this?

There are websites[3] you can visit to test your face recognition skills, so you can find some evidence on whether your face recognition skills are within the normal range or not. However, it is strongly recommended that you contact a specialist in your area for a complete assessment of your memory, perception and so forth. If you cannot find the relevant information online, I invite you to send me an email (prosoricerca@googlemail.com) and I will be happy to help you in finding the right specialist.

If you think you are prosopagnosic, do not excessively worry; as mentioned many times in the book, it does not mean that you have a psychiatric or neurological disorder and by no means indicates that you have somehow impaired cognitive or intellective functioning.

I advise you to inform other people regarding your problems; tell them that if it happens that you do not say hallo it is not because you are rude, but that you might suffer from a problem in identifying faces. Say this to you relatives, family, neighbors, work colleagues; this will improve you social relations and will augment your confidence in interpersonal situations. Please do not use this as an excuse when you do not want to greet someone you do not like!

References

Ariel, R., & Sadeh, M. (1996). Congenital visual agnosia and prosopagnosia in a child: A case report. *Cortex, 32,* 221–240.

Behrmann, M., Avidan, G., Marotta, J. J., & Kimchi, R. (2005). Detailed exploration of face-related processing in congenital prosopagnosia: 1. *Behavioral Findings. Journal of Cognitive Neuroscience, 17,* 1130–1149.

Brunsdon, Ruth, Coltheart, Max, Nickels, Lyndsey, & Joy, Pamela. (2006). Developmental prosopagnosia: A case analysis and treatment study. *Cognitive Neuropsychology, 23,* 822–840.

Duchaine, B., Germine, L., & Nakayama, K. (2007). Family resemblance: Ten family members with prosopagnosia and within-class object agnosia. *Cognitive Neuropsychology, 24,* 419–430.

---

[3] http://www.faceblind.org/facetests/

Duchaine, B., & Nakayama, K. (2005). Dissociations of face and object recognition in developmental prosopagnosia. *Journal of Cognitive Neuroscience, 17*, 249–261.

Elgar, K., & Campbell, R. (2001). Annotation: The cognitive neuroscience of face recognition: Implications for developmental disorders. *Journal of Child Psychology and Psychiatry, 42*, 705–717.

Honey, E., McConachie, H., Randle, V., Shearer, H., & Le Couteur, A. S. (2008). One-year change in repetitive behaviours in young children with communication disorders including autism. *Journal of Autism and Developmental Disorders, 38*, 1439–1450.

Jones, R. D., & Tranel, D. (2001). Severe developmental prosopagnosia in a child with superior intellect. *Journal of Clinical and Experimental Neuropsychology, 23*, 265–273.

Klin, A., Jones, W., Schultz, R. T., Volkmar, F., & Cohen, D. (2002). Visual fixation pattern during viewing of naturalistic social situations as predictors of social competence in individuals with autism. *Archives of General Psychiatry, 59*, 809–816.

Kracke, I. (1994). Developmental prosopagnosia in Asperger syndrome: Presentation and discussion of an individual case. *Developmental Medicine abd Child Neurology, 36*, 873–886.

McConachie, H. R. (1976). Developmental prosopagnosia: A single case report. *Cortex, 12*, 76–82.

Moore, V., & Goodson, S. (2003). How well does early diagnosis of autism stand the test of time. *Autism, 7*, 47–63.

Nelson, C. A. (2001). The development and neural bases of face recognition. *Infant and Child Development, 10*, 3–18.

Schmalzl, L., Palermo, R., Green, M., Brunsdon, R., & Coltheart, M. (2008). Training of familiar face recognition and visual scan paths for faces in a child with congenital prosopagnosia. *Cognitive Neuropsychology, 25*, 704–729.

Schultz, R. T. (2005). Developmental deficits in social perception in autism: The role of the amygdala and fusiform face area. *International Journal of Developmental Neuroscience, 23*, 125–141.

Tippett, L. J., Miller, L. A., & Farah, M. J. (2000). Prosopamnesia: A selective impairment in face learning. *Cognitive Neuropsychology, 17*, 241–255.

Williams, M., Berberovic, N., & Mattingley, J. (2007). Abnormal fMRI adaptation to unfamiliar faces in a case of developmental prosopamnesia. *Current Biology, 17*, 1259–1264.

Wilson, C. E., Brock, J., & Palermo, R. (2011). Recognition of own- and other-race faces in autism spectrum disorders. *Quarterly Journal of Experimental Psychology, 64*, 1939–1954.

Wilson, C. E., Palermo, R., & Brock, J. (2012). Visual scan paths and recognition of facial identity in autism spectrum disorder and typical development. *PLoS ONE, 7*, e37681.

Printed by Publishers' Graphics LLC
MLSI130930.15.16.649